SCHAUM'S *Easy* OUTLINES

ELECTRIC CIRCUITS

Other Books in Schaum's Easy Outlines Series Include:

Schaum's Easy Outline: Calculus
Schaum's Easy Outline: College Algebra
Schaum's Easy Outline: College Mathematics
Schaum's Easy Outline: Differential Equations
Schaum's Easy Outline: Discrete Mathematics
Schaum's Easy Outline: Elementary Algebra
Schaum's Easy Outline: Geometry
Schaum's Easy Outline: Intermediate Algebra
Schaum's Easy Outline: Linear Algebra
Schaum's Easy Outline: Mathematical Handbook of Formulas and Tables
Schaum's Easy Outline: Precalculus
Schaum's Easy Outline: Probability and Statistics
Schaum's Easy Outline: Statistics
Schaum's Easy Outline: Trigonometry
Schaum's Easy Outline: Bookkeeping and Accounting
Schaum's Easy Outline: Business Statistics
Schaum's Easy Outline: Economics
Schaum's Easy Outline: Principles of Accounting
Schaum's Easy Outline: Beginning Chemistry
Schaum's Easy Outline: Biology
Schaum's Easy Outline: Biochemistry
Schaum's Easy Outline: College Chemistry
Schaum's Easy Outline: Genetics
Schaum's Easy Outline: Human Anatomy and Physiology
Schaum's Easy Outline: Molecular and Cell Biology
Schaum's Easy Outline: Organic Chemistry
Schaum's Easy Outline: Applied Physics
Schaum's Easy Outline: Physics
Schaum's Easy Outline: HTML
Schaum's Easy Outline: XML
Schaum's Easy Outline: Physics
Schaum's Easy Outline: Programming with C++
Schaum's Easy Outline: Programming with Java
Schaum's Easy Outline: Basic Electricity
Schaum's Easy Outline: Electromagnetics
Schaum's Easy Outline: Introduction to Psychology
Schaum's Easy Outline: French
Schaum's Easy Outline: German
Schaum's Easy Outline: Italian
Schaum's Easy Outline: Spanish
Schaum's Easy Outline: Writing and Grammar

SCHAUM'S *Easy* OUTLINES

ELECTRIC CIRCUITS

BASED ON SCHAUM'S
Outline of Theory and Problems of Electric Circuits
BY

MAHMOOD NAHVI, Ph.D.
AND
JOSEPH A. EDMINISTER

ABRIDGEMENT EDITOR
WILLIAM T. SMITH, Ph.D.

SCHAUM'S OUTLINE SERIES

McGRAW-HILL

New York Chicago San Francisco Lisbon London Madrid Mexico City Milan New Delhi San Juan Seoul Singapore Sydney Toronto

The McGraw-Hill Companies

MAHMOOD NAHVI is Professor of Electrical Engineering at California Polytechnic State University in San Luis Obispo, California. He earned B.Sc., M.Sc., and Ph.D. degrees in electrical engineering and has 44 years of teaching and research experience. His areas of special interest include network theory, control theory, communications engineering, signal processing, neural networks, adaptive control and learning in synthetic and living systems, communication and control in the central nervous system, and engineering education. In the area of engineering education, he has developed computer modules for electric circuits, signals, and systems that improve teaching and learning of the fundamentals of electrical engineering.

JOSEPH A. EDMINISTER is Professor Emeritus of Electrical Engineering from the University of Akron in Ohio, where he taught from 1957 until his retirement in 1983. He was also Assistant Dean and Acting Dean of Engineering. After serving on the staff of Ohio Congressman Dennis Eckart in 1984, Edminister joined Cornell University as a patent attorney and later as Director of Corporate Relations for the College of Engineering. He retired from Cornell in 1995. He received a bachelor of science in electrical engineering, a master of science in engineering, and a juris doctorate from the University of Akron. He is a registered Professional Engineer in Ohio as well as a member of the bar and a registered patent attorney. He is the author of *Schaum's Outline of Theory and Problems of Electromagnetics.*

WILLIAM T. SMITH is Associate Professor in the department of electrical engineering at the University of Kentucky, where he has taught since 1990 and has twice won the Outstanding Engineering Professor Award. He earned a B.S. degree from the University of Kentucky and both M.S. and Ph.D. degrees in electrical engineering from the Virginia Polytechnic Institute and State University. He was an academic visitor at the IBM Austin Research Laboratory and previously worked as a senior engineer in the Government Aerospace Systems Division of Harris Corporation. He is also the co-author of several journal articles and conference papers and served as abridgement editor of *Schaum's Easy Outline: Basic Electricity* and *Schaum's Easy Outline: Electromagnetics.*

2 3 4 5 6 7 8 9 0 DOC DOC 0 9 8 7 6 5 4

ISBN 0-07-142241-2

Contents

Chapter 1
CIRCUIT CONCEPTS

IN THIS CHAPTER:

- ✔ *Electrical Quantities*
- ✔ *Passive and Active Elements*
- ✔ *Resistance, Inductance, and Capacitance*
- ✔ *Circuit Diagrams*
- ✔ *Nonlinear Resistors*

Electrical Quantities

The International System of Units (SI) will be used throughout this book. Four of the basic quantities and their SI units are listed in Table 1-1. Three other basic quantities are temperature (degrees kelvin, K), amount of substance (moles, mol), and luminous intensity (candelas, cd). All other units may be derived from the seven basic units. Force is measured in *newtons* (N). 1 N = 1kg · m/s^2. Work is measured in force over distance using *joules* (J). 1 J = 1 N · m. Power is the rate at which work is done and is measured in *watts* (W). 1 W = 1 J/s.

The electrical quantities and their symbols commonly used in elec-

Table 1-1

Quantity	Symbol	SI Unit	Abbreviation
length	L, l	meter	m
mass	M, m	kilogram	kg
time	T, t	second	s
current	I, i	ampere	A

trical circuit analysis are listed in Table 1-2. The decimal multiples or submultiples of SI units should be used whenever possible. The symbols given in Table 1-3 are prefixed to the unit symbols of Tables 1-1 and 1-2.

For letters used as symbols, non-time varying quantities are capitalized and time-varying quantities are lower case.

Table 1-2

Quantity	Symbol	SI Unit	Abbreviation
electric charge	Q, q	coulomb	C
electric potential	V, v	volt	V
resistance	R	ohm	Ω
conductance	G	siemens	S
inductance	L	henry	H
capacitance	C	farad	F
frequency	f	hertz	Hz
force	F, f	newton	N
energy, work	W, w	joule	J
power	P, p	watt	W
magnetic flux	ϕ	weber	Wb
magnetic flux density	$\mathbf{B}$	tesla	T

Table 1-3

Prefix	Factor	Symbol
pico	10^{-12}	p
nano	10^{-9}	n
micro	10^{-6}	μ
milli	10^{-3}	m
centi	10^{-2}	c
deci	10^{-1}	d
kilo	10^{3}	k
mega	10^{6}	M
giga	10^{9}	G
tera	10^{12}	T

Electric charge is measured in *coulombs* (C). Electric current is electric charge in motion and is measured in *amperes* (A). 1 A = 1 C/s. The common symbols for current are I, i. Positive charges moving to the left, as suggested in Figure 1-1(a) produce a current i, directed to the left. Negative charges moving to the right as shown in Figure 1-1(b) also produce a current directed to the left.

Electrons are the basic unit of electric charge and have a value of 1602×10^{-19} C.

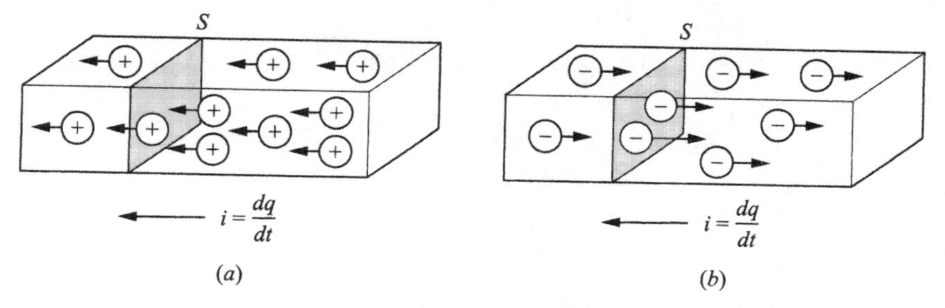

Figure 1-1 Current flow versus direction of charge movement

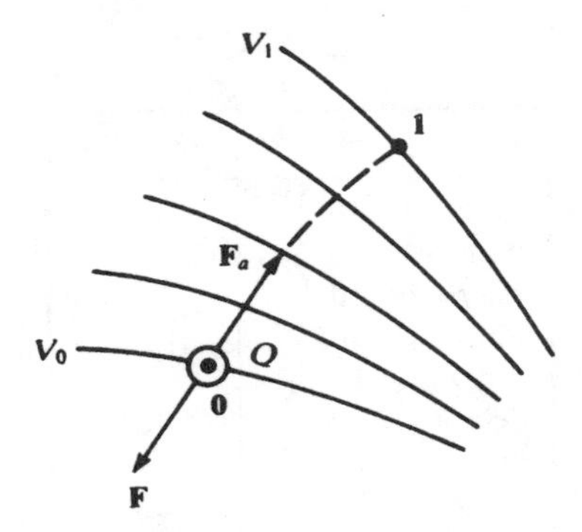

Figure 1-2 Illustrating the definition of potential

Electric current in metallic conductors takes place through the motion of electrons that occupy the outermost shell of the atomic structure.

An electric charge experiences a force in an electric field, which if unopposed, will accelerate the particle containing the charge. Of interest here is the work done to move the charge against the field as suggested in Figure 1-2.

If 1 joule of work is required to move the charge Q of 1 coulomb from position 0 to position 1, then position 1 is at a potential of 1 *volt* with respect to position 0. $1\text{V} = 1\text{J/C}$. Potential is commonly referred to as *voltage* and V,v are often used as symbols to represent voltage.

Electric energy is stored in electric and magnetic fields.

The rate in joules at which energy is transferred is electric power in *watts* (W).

Furthermore, the product of voltage and current yields the electric power; $p = vi$; $1\text{W} = 1\text{V} \cdot \text{A}$. In a more fundamental sense, power is the time derivative of energy $p = dw/dt$.

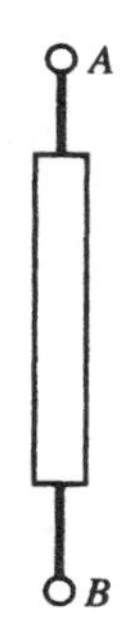

Figure 1-3 General circuit element

Passive and Active Elements

An electrical device is often represented by a *circuit diagram* or a *network* constructed from series and parallel arrangement of two-terminal elements. A two-terminal element in general form is shown in Figure 1-3 with a single device represented by the rectangle and two perfectly conducting leads ending at connecting points A and B.

Active elements are voltage or current sources that are able to supply energy to the circuit. Resistors, inductors, and capacitors are examples of *passive* circuits that take energy from the sources and either convert it to another form or store it in an electric or magnetic field.

Figure 1-4 illustrates seven basic circuit elements.

Elements (a) and (b) are voltage sources and (c) and (d) are current sources. A voltage source that is not affected by changes in the connected circuit is an *independent* source, illustrated by the circle in Figure 1-4(a). A *dependent* voltage source that changes in some described man-

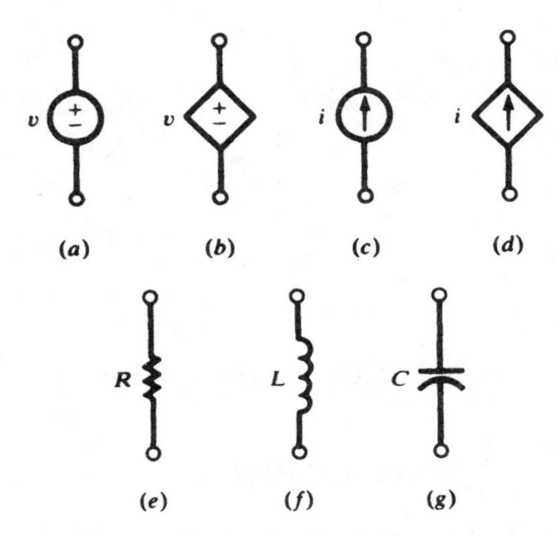

Figure 1-4 Basic circuit elements

ner with the conditions on the connected circuit is shown by the diamond-shaped symbol in Figure 1-4(*b*). Current sources may also be either independent or dependent and the corresponding symbols are shown in (*c*) and (*d*). The passive resistor, inductor, and capacitor are shown in Figures (*e*)–(*g*), respectively.

A voltage function and a polarity must be specified to completely describe a voltage source. In Figure 1-5(*a*), the polarity marks, + and –, are placed near the conductors of the symbol that identifies the voltage source.

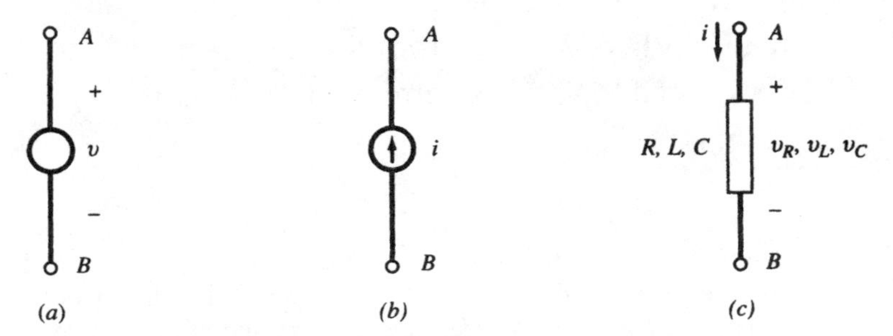

Figure 1-5 Polarity and current flow in devices

Similarly, a current source requires that a direction be indicated as shown in Figure 1-5(*b*). For passive circuit elements, shown in Figure 1-5(*c*), the terminal where the current enters is generally treated as positive with respect to the terminal where the current leaves.

Remember

Power is absorbed by a device if current enters the positive terminal. Power is generated if current leaves the positive terminal. Power is computed using $P = VI = I^2R = V^2/R$.

The dc (dc meaning non-time varying) circuit of Figure 1-6 is used to illustrate the power calculation. The $5-\Omega$ resistor absorbs $(3A)^2(5\Omega) = 45$ W and the 5-V battery V_B absorbs (3A)(5V) = 15 W of power. The 20-V battery V_A generates (3A)(20V) = 60 W of power. In this *conservative* circuit, the total generated power equals the total absorbed power.

Resistance, Inductance, and Capacitance

The passive circuit elements resistors *R*, inductors *L*, and capacitors *C* are defined by the manner in which the voltage and current are related for the individual element. Table 1-4 summarizes these relationships for the

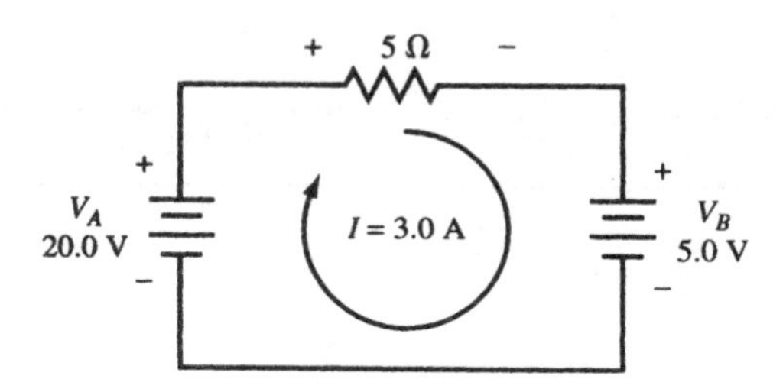

Figure 1-6 Circuit illustrating power calculation

Table 1-4

Circuit element	Units	Voltage	Current	Power
i, $+$, v, $-$ **Resistance**	ohms (Ω)	$v = Ri$ **(Ohms's law)**	$i = \frac{v}{R}$	$p = vi = i^2 R$
i, $+$, v, $-$ **Inductance**	henries (H)	$v = L\frac{di}{dt}$	$i = \frac{1}{L}\int v\,dt + k_1$	$p = vi = Li\frac{di}{dt}$
i, $+$, v, $-$ **Capacitance**	farads (F)	$v = \frac{1}{C}\int i\,dt + k_2$	$i = C\frac{dv}{dt}$	$p = vi = Cv\frac{dv}{dt}$

three passive circuit elements. Note the current directions and the corresponding polarity of the voltages.

All devices that consume energy must have a resistor (also called a *resistance*) in their circuit model. Power in the resistor is given by

$$p = vi = i^2R = v^2/R$$

which is always positive. Energy is determined as the integral of the instantaneous power

$$w = \int_{t_1}^{t_2} p\,dt = R\int_{t_1}^{t_2} i^2\,dt = \frac{1}{R}\int_{t_1}^{t_2} v^2\,dt$$

Example 1.1 A $4-\Omega$ resistor has a current $i = 2.5\sin\omega t$ (A). Find the voltage, power and the energy over one cycle.

Solution:

$$v = Ri = 10\sin\omega t \text{ (V)}$$
$$p = vi = i^2R = 25\sin^2\omega t \text{ (W)}$$
$$w = \int_0^t p\,dt = 25\left[\frac{t}{2} - \frac{\sin 2\omega t}{4\omega}\right] \text{ (J)}$$

The plots in Figure 1-7 illustrate that p is always positive and that the energy absorbed by the resistor is always increasing.

The circuit element that stores energy in a magnetic field is an inductor (also called an *inductance*). With time-variable current, the energy is generally stored during some parts of the cycle and then returned to the source during others. Inductors are often constructed using coils. The power and energy relationships are as follows:

$$p = vi = L\frac{di}{dt}i = \frac{d}{dt}\left[\frac{1}{2}Li^2\right]$$
$$w_L = \int_{t_1}^{t_2} p\,dt = \frac{1}{2}L\left[i_2^2 - i_1^2\right]$$

Example 1.2 In the interval $0 < t < (\pi/50)$ s, a 30-mH inductance has a current $i = 10\sin 50t$ (A). Obtain the voltage, power, and energy for the inductance.

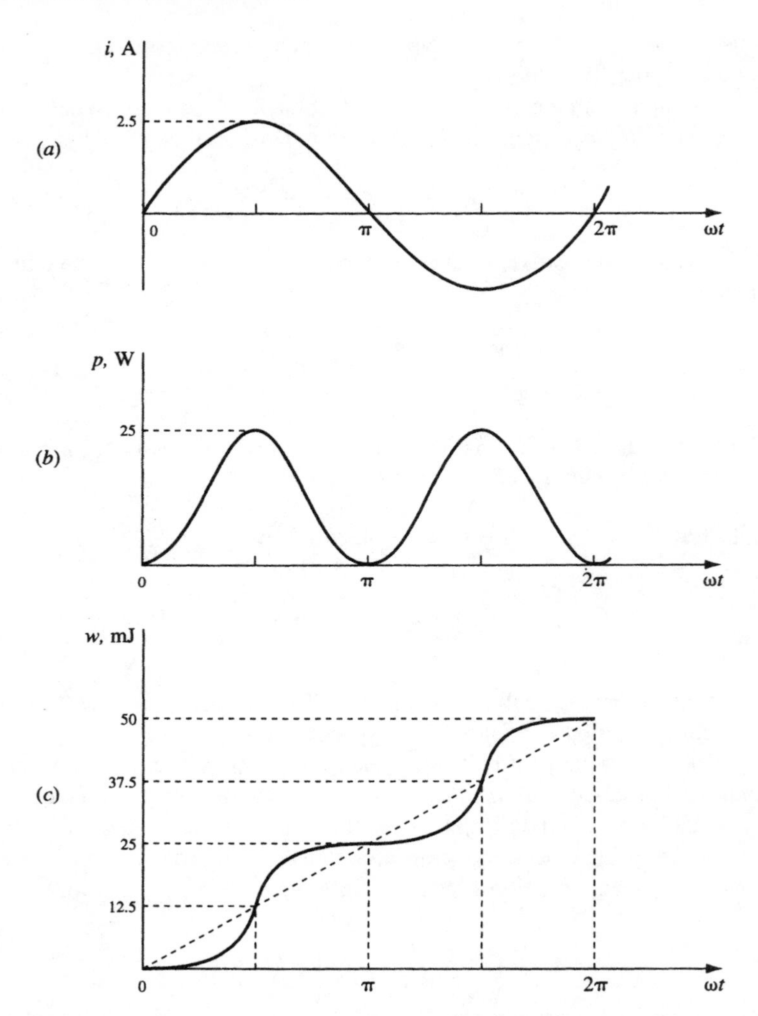

Figure 1-7 Plots of current, power, and energy for the resistor

Solution:

$$v = L\frac{di}{dt} = 15\cos 50t \text{ (V)}$$

$$p = vi = 75\sin 100t \text{ (W)}$$

$$w_L = \int_{t_1}^{t_2} p\,dt = 0.75(1-\cos 100t) \text{ (J)}$$

As shown in Figure 1-8, the energy is zero at $t = 0$ and $t = \pi/50$ s. Thus, while energy transfer did occur over the interval, this energy was first stored and later returned to the source.

The circuit element that stores energy in an electric field is a capacitor (also called a *capacitance*). When the voltage is variable over a cycle, energy will be stored during one part of the cycle and returned in the next. While an inductance cannot retain energy after removal of the source because the magnetic field collapses, the capacitor retains the charge and the electric field can remain after the source is removed. This charged condition can remain until a discharge path is provided, at which time the energy is released. The power and energy relationships for the capacitance are as follows:

$$p = vi = Cv\frac{dv}{dt} = \frac{d}{dt}\left[\frac{1}{2}Cv^2\right]$$

$$w_C = \int_{t_1}^{t_2} p\,dt = \frac{1}{2}C\left[v_2^2 - v_1^2\right]$$

Example 1.3 In the interval $0 < t < 5\pi$ ms, a 20-μF capacitance has a voltage $v = 50\sin 200t$ (V). Obtain the current, power and energy. Plot w_C assuming $w = 0$ at $t = 0$.

Solution:

$$i = C\frac{dv}{dt} = 0.2\cos 200t \text{ (A)}$$

$$p = vi = 5\sin 400t \text{ (W)}$$

$$w_C = \int_{t_1}^{t_2} p\,dt = 12.5[1 - \cos 400t]\text{(mJ)}$$

Figure 1-9 shows that the stored energy increases to a value of 25 mJ, after which it returns to zero as the energy is returned to the source.

Circuit Diagrams

Every circuit diagram can be constructed in a variety of ways that may look different but are in fact identical. The diagram presented in a problem may not suggest the best of several methods of solution. Consequently, a diagram should be examined before a solution is started and re-

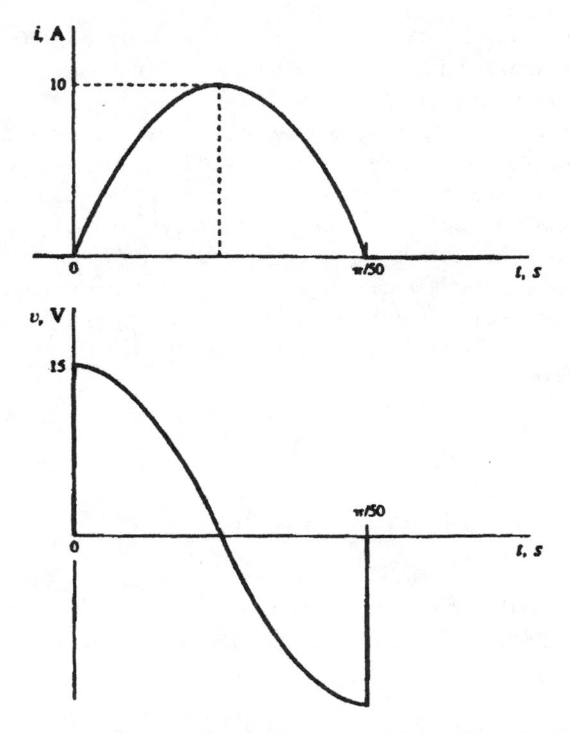

Figure 1-8 Power and energy for an inductor

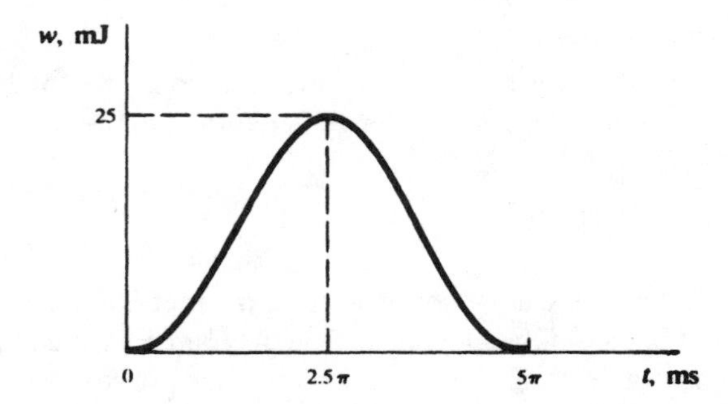

Figure 1-9 Energy storage in a capacitor

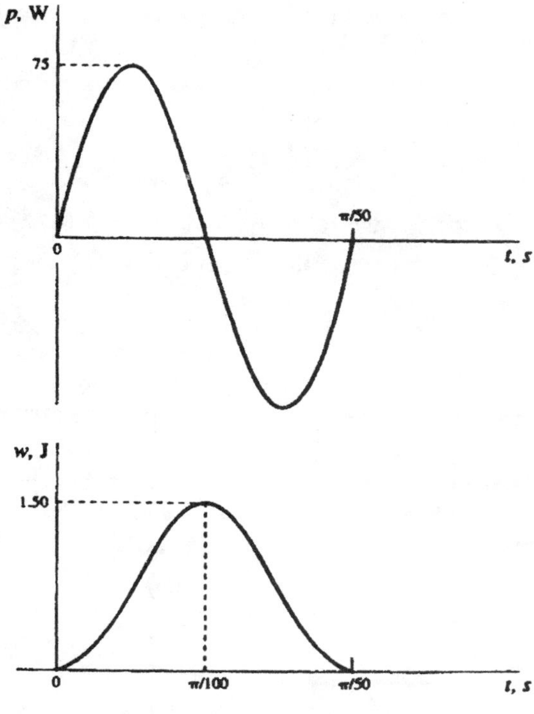

Figure 1-8, continued

drawn if necessary to show more clearly how the elements are interconnected. An extreme example is illustrated in Figure 1-10 where the circuits are actually functionally identical. In Figure 1-10(*a*), the three "junctions" labeled *A* are shown as two "junctions" in (*b*). However, resistor R_4 is bypassed by a short circuit and may be removed for purposes of analysis. Then, in Figure 1-10(*c*), the single junction *A* is shown with its three meeting branches.

Nonlinear Resistors

The current-voltage relationship in an element may be instantaneous but not necessarily linear. The element is then modeled as a nonlinear resistor. An example is a filament lamp, which at higher voltages draws proportionally less current. Another important example of a nonlinear resistor is a diode.

Note!

A diode is a two-terminal device that, roughly speaking, conducts electric current in one direction.

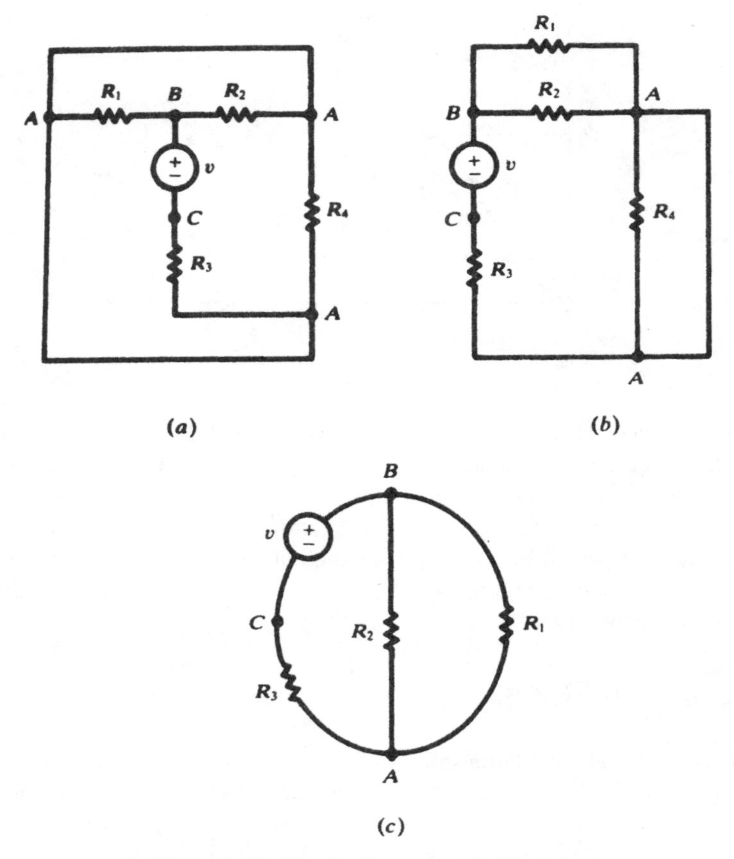

Figure 1-10 Equivalent circuit diagrams

The *static resistance* of a nonlinear resistor operating at (I,V) is $R = V/I$. Its *dynamic resistance* is $r = \Delta V/\Delta I$ which is the inverse slope of the current plotted versus voltage. Static and dynamic resistances both depend on the operating point.

Example 1.4 The current and voltage characteristic of a semiconductor diode in the forward direction is measured and recorded in the following table:

v(V)	0.5	0.6	0.65	0.66	0.67	0.68	0.69	0.7	0.71	0.72	0.73	0.74	0.75
i(mA)	$2\text{x}10^{-4}$	0.11	0.78	1.2	1.7	2.6	3.9	5.8	8.6	12.9	19.2	28.7	42.7

In the reverse direction (i.e., when $v < 0$), $I = 4 \times 10^{-15}$ (A). Using the values given in the table, calculate the static and dynamic resistances (R and r) of the diode when it operates at 30 mA and find its power consumption.

Solution: From the table,

$$R = \frac{V}{I} \approx \frac{0.74}{28.7 \times 10^{-3}} = 25.78\ \Omega$$

$$r = \frac{\Delta V}{\Delta I} \approx \frac{0.75 - 0.73}{(42.7 - 19.2) \times 10^{-3}} = 0.85\ \Omega$$

$$p = vi \approx 0.74 \times 28.7 \times 10^{-3} = 21.238\,\text{mW}$$

Important Things to Remember

- ✔ Electric current is electric charge in motion. 1 A of current is equivalent to 1 C of charge moving across a fixed surface in 1 s.
- ✔ Power is the rate at which energy is transferred. 1 W of power is equivalent to 1 J transferred in 1 s.
- ✔ Active circuit elements supply energy to a circuit.
- ✔ Passive elements either convert energy to a different form or store it in an electric or magnetic field.
- ✔ The power in a circuit element is positive if the current flowing through enters the positive potential terminal.
- ✔ The relationship between voltage and current in a resistor is algebraic.
- ✔ The relationship between voltage and current in an inductor or capacitor is calculus-based.
- ✔ Circuit diagrams for the same circuit can look dramatically different.
- ✔ Nonlinear resistors are often characterized by the plot of i versus v.

Chapter 2
Circuit Analysis Methods

In This Chapter:

- ✔ *Kirchhoff's Voltage Law*
- ✔ *Kirchhoff's Current Law*
- ✔ *Series Circuits and Voltage Division*
- ✔ *Parallel Circuits and Current Division*
- ✔ *Mesh Current Method*
- ✔ *Node Voltage Method*
- ✔ *Network Reduction*
- ✔ *Superposition*
- ✔ *Thévenin's and Norton's Theorems*
- ✔ *Maximum Power Transfer*

Kirchhoff's Voltage Law

For any closed path in a network, *Kirchhoff's voltage law (KVL)* states that the algebraic sum of the voltages is zero. Some of the voltages will be sources while others will result from current in passive elements creating a voltage. This is sometimes referred to as a *voltage drop*.

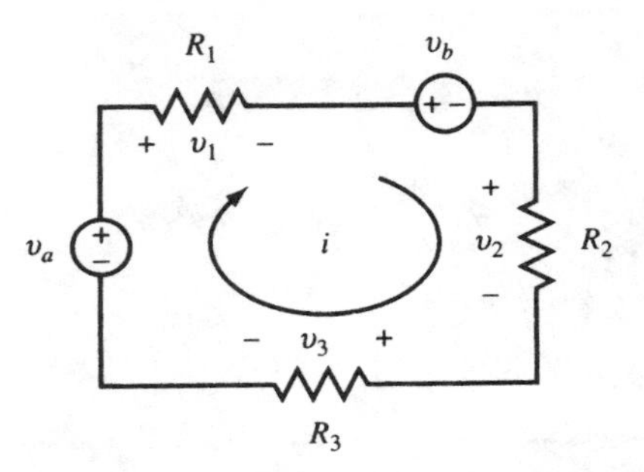

Figure 2-1 Example circuit for Kirchhoff's voltage law

Example 2.1 Write the KVL equation for the circuit shown in Figure 2-1.

Solution: Starting at the lower left corner of the circuit, for the current direction as shown, we have

$$-v_a + v_1 + v_b + v_2 + v_3 = 0$$
$$-v_a + iR_1 + v_b + iR_2 + iR_3 = 0$$
$$v_a - v_b = i(R_1 + R_2 + R_3)$$

Kirchhoff's Current Law

The connection of two or more circuit elements creates a junction called a *node*. The junction between two elements is called a *simple node*. The junction of three or more elements is called a *principal node*.

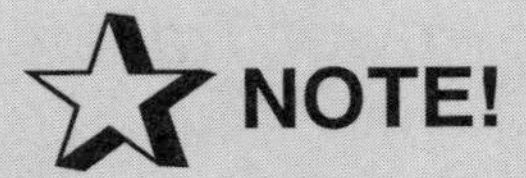

At a principal node, current division does take place.

Kirchhoff's current law (KCL) states that the algebraic sum of the currents entering a node is equal to the sum of the currents leaving that node. The basis for the law is conservation of electric charge.

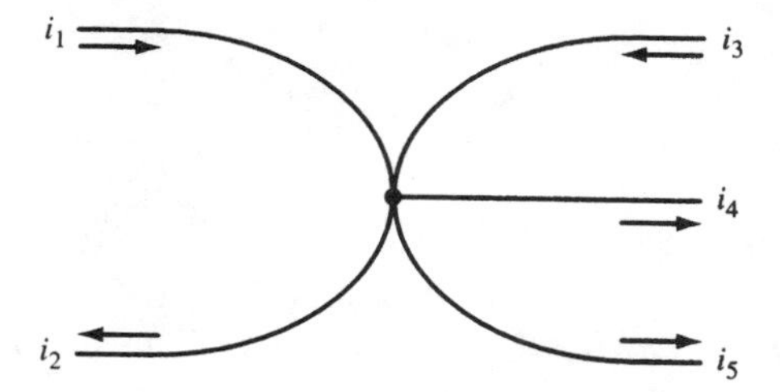

Figure 2-2 Example circuit for Kirchhoff's current law

Example 2.2 Write the KCL equation for the principal node shown in Figure 2-2.

Solution:

$$i_1 - i_2 + i_3 - i_4 - i_5 = 0$$
$$i_1 + i_3 = i_2 + i_4 + i_5$$

Series Circuits and Voltage Division

Three passive circuit elements in series connections as shown in Figure 2-3 have the same current i. The voltages across the elements are v_1, v_2, and v_3. The total voltage v is the sum of the individual voltages:

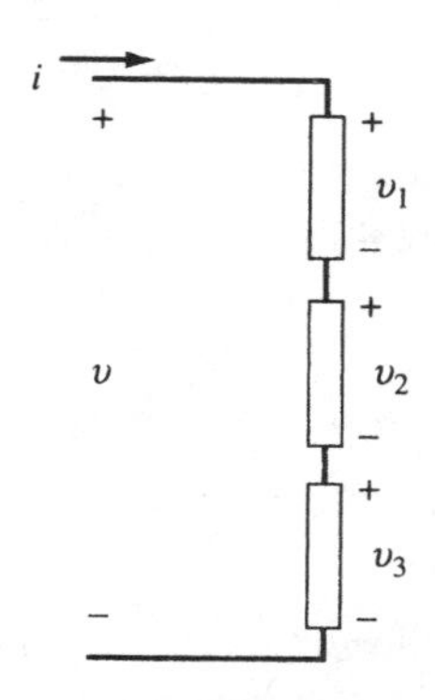

Figure 2-3 Three elements connected in series

$$
\begin{aligned}
v &= v_1 + v_2 + v_3 \\
v &= iR_1 + iR_2 + iR_3 \\
v &= i(R_1 + R_2 + R_3) \\
v &= iR_{eq}
\end{aligned}
$$

where $R_{eq} = R_1 + R_2 + R_3$. If the three passive elements are inductances,

$$
\begin{aligned}
v &= L_1 \frac{di}{dt} + L_2 \frac{di}{dt} + L_3 \frac{di}{dt} \\
v &= (L_1 + L_2 + L_3) \frac{di}{dt}
\end{aligned}
$$

The equivalent series inductance would be $L_{eq} = L_1 + L_2 + L_3$
If the three circuit elements are capacitances, assuming zero initial charges so the constants of integration are zero,

$$
\begin{aligned}
v &= \frac{1}{C_1} \int i\,dt + \frac{1}{C_2} \int i\,dt + \frac{1}{C_3} \int i\,dt \\
v &= \left(\frac{1}{C_1} + \frac{1}{C_2} + \frac{1}{C_3} \right) \int i\,dt
\end{aligned}
$$

The equivalent series capacitance would be found using

$$
\frac{1}{C_{eq}} = \left(\frac{1}{C_1} + \frac{1}{C_2} + \frac{1}{C_3} \right)
$$

Example 2.3 The equivalent resistance of three resistors in series is 750 Ω. Two of the resistors are 40 and 410 Ω. What is the value of the third resistor?

Solution:

$$
\begin{aligned}
R_{eq} &= R_1 + R_2 + R_3 \\
R_3 &= R_{eq} - R_1 - R_2 = 300\ \Omega
\end{aligned}
$$

Example 2.4 Two capacitors, $C_1 = 2\ \mu\text{F}$ and $C_2 = 10\ \mu\text{F}$ are connected in series. Find the equivalent capacitance.

Solution:
$$\frac{1}{C_{eq}} = \left(\frac{1}{C_1} + \frac{1}{C_2}\right)$$

Using algebra,

$$C_{eq} = \frac{C_1 C_2}{C_1 + C_2} = \frac{(2\times10^{-6})(10\times10^{-6})}{2\times10^{-6} + 10\times10^{-6}} = 1.67\ \mu\text{F}$$

A set of series-connected resistors as shown in Figure 2-4 is referred to as a *voltage divider*.

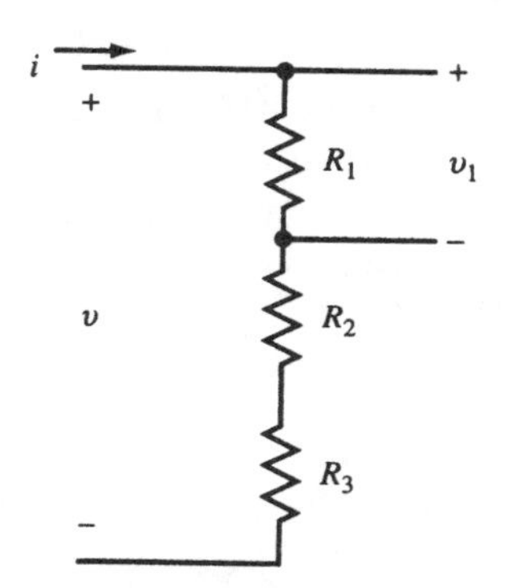

Figure 2-4 Voltage divider

Since $v_1 = iR_1$ and $v = i(R_1 + R_2 + R_3)$,

$$v_1 = v\left(\frac{R_1}{R_1 + R_2 + R_3}\right)$$

The voltage v divides across the resistors according to the resistance ratio $\frac{R_1}{R_1 + R_2 + R_3}$ for v_1, and similarly for v_2 and v_3.

Example 2.5 A voltage divider circuit of two resistors is designed with a total resistance of the two resistors equal to 50 Ω. If the output voltage is 10 percent of the input voltage, obtain the values of the two resistors in the circuit.

Solution: The voltage ratio is

$$\frac{v_1}{v} = 0.10 = \frac{R_1}{R_{eq}} = \frac{R_1}{50}$$

Therefore, $R_1 = 5\ \Omega$ and from $R_{eq} = R_1 + R_2$, $R_2 = R_{eq} - R_1 = 45\ \Omega$

Parallel Circuits and Current Division

For three circuits connected in parallel as shown in Figure 2-5, KCL states that the current i entering the principal node is the sum of the three currents leaving the node through the branches. If the three passive elements are resistances,

$$i = \frac{v}{R_1} + \frac{v}{R_2} + \frac{v}{R_3} = \left(\frac{1}{R_1} + \frac{1}{R_2} + \frac{1}{R_3}\right)v = \frac{1}{R_{eq}}v$$

For the three resistances in parallel,

$$\frac{1}{R_{eq}} = \frac{1}{R_1} + \frac{1}{R_2} + \frac{1}{R_3}$$

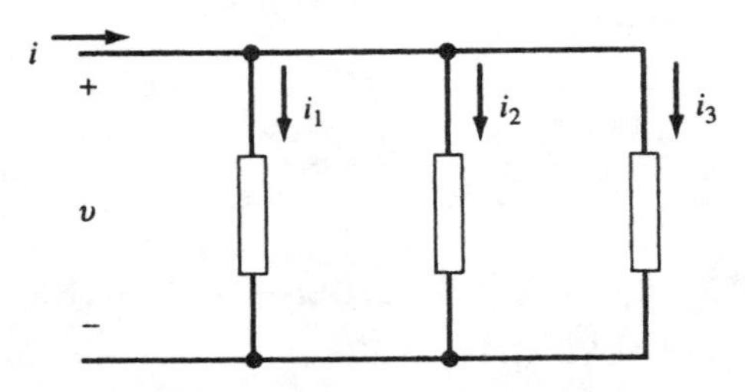

Figure 2-5 Three circuits connected in parallel

You Need to Know 

The case of two resistances in parallel occurs frequently and deserves special mention. Using algebra, the equivalent resistance is given by the *product over the sum*

$$R_{eq} = \frac{R_1 R_2}{R_1 + R_2}$$

Example 2.6 Obtain the equivalent resistance of (*a*) two 60 Ω resistors in parallel and (*b*) three 60 Ω resistors in parallel.

Solution: (*a*) $R_{eq} = \frac{(60)^2}{120} = 30\ \Omega$

(*b*) $\frac{1}{R_{eq}} = \frac{1}{60} + \frac{1}{60} + \frac{1}{60} \quad \Rightarrow \quad R_{eq} = 20\ \Omega$

For *n* identical resistors *R* in parallel, $R_{eq} = R/n$.

Combinations of inductances in parallel have similar expressions to those of resistors in parallel. The equivalent inductance for three inductances in parallel is given by

$$\frac{1}{L_{eq}} = \frac{1}{L_1} + \frac{1}{L_2} + \frac{1}{L_3}$$

and for two inductances in parallel

$$L_{eq} = \frac{L_1 L_2}{L_1 + L_2}$$

Example 2.7 Two inductances $L_1 = 3$ mH and $L_2 = 6$ mH are connected in parallel. Find L_{eq}.

Solution:
$$L_{eq} = \frac{(3\times10^{-3})(6\times10^{-3})}{3\times10^{-3}+6\times10^{-3}} = 2 \text{ mH}$$

With three capacitances in parallel,

$$i = C_1\frac{dv}{dt} + C_2\frac{dv}{dt} + C_3\frac{dv}{dt} = (C_1 + C_2 + C_3)\frac{dv}{dt} = C_{eq}\frac{dv}{dt}$$

The equivalent capacitance for three capacitors in parallel is $C_{eq} = C_1 + C_2 + C_3$ which is of the same form as resistors in series.

A parallel arrangement of resistors as shown in Figure 2-6 results in a *current divider*. The ratio of branch current i_1 to the total current i illustrates the operation of the divider. Since $i_1 = \frac{v}{R_1}$ and the total current is $i = \frac{v}{R_1} + \frac{v}{R_2} + \frac{v}{R_3}$,

$$\frac{i_1}{i} = \frac{1/R_1}{1/R_1 + 1/R_2 + 1/R_3}$$

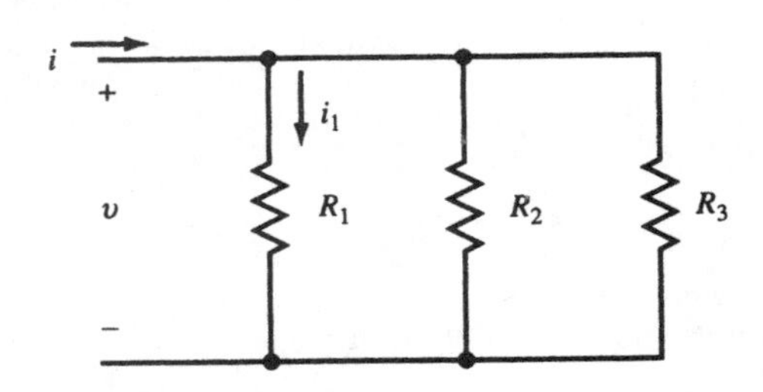

Figure 2-6 Current divider

For a two-branch divider, we have

$$\frac{i_1}{i} = \frac{R_2}{R_1 + R_2}$$

> The ratio of the current in one branch of a two-branch parallel circuit to the total current is equal to the ratio of the resistance of the *other* branch resistance to the sum of the two resistances.

Example 2.8 A current of 30 mA is to be divided into two branch currents of 20 mA and 10 mA by a network with an equivalent resistance equal to 10 Ω. Obtain the branch resistances.

Solution: From current division,

$$\frac{20}{30} = \frac{R_2}{R_1 + R_2} \text{ and } \frac{10}{30} = \frac{R_1}{R_1 + R_2}. \text{ Also, } R_{eq} = \frac{R_1 R_2}{R_1 + R_2} = 10\ \Omega$$

Solving these equations gives $R_1 = 15\Omega$ and $R_2 = 30\ \Omega$.

Mesh Current Method

In the mesh current method, a current is assigned to each *window* of the network such that the currents complete a closed loop. They are referred to as *mesh* or *loop* currents. Each element and branch therefore will have an independent current. When a branch has two of the mesh currents, the actual current is given by their algebraic sum. The assigned mesh currents may have either clockwise or counterclockwise direction, although at the outset it is wise to assign to all of the mesh currents a clockwise direction. Once the currents are assigned, Kirchhoff's voltage law is written for each loop to obtain the necessary simultaneous equations.

Example 2.9 Obtain the current in each branch of the network shown in Figure 2-7 using the mesh current method.

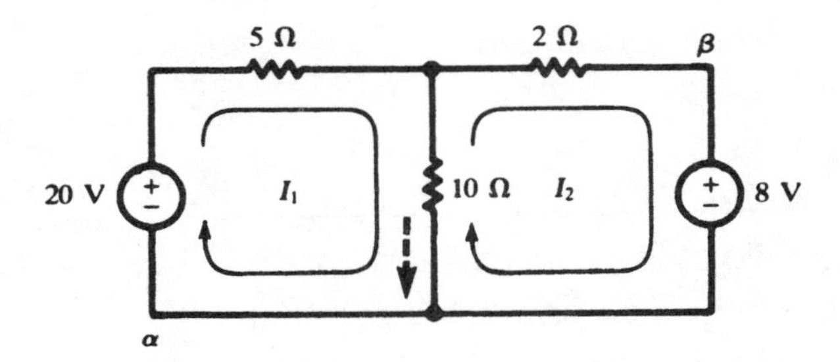

Figure 2-7 Mesh current example circuit

Solution: The currents I_1 and I_2 are chosen as shown on the circuit diagram. Applying KVL around the left loop, starting at point α,

$$-20 + 5I_1 + 10(I_1 - I_2) = 0$$

and around the right loop, starting at point β,

$$8 + 10(I_2 - I_1) + 2I_2 = 0$$

Rearranging terms,

$$15I_1 - 10I_2 = 20$$
$$-10I_1 + 12I_2 = -8$$

Solving the equations simultaneously results in $I_1 = 2$ A and $I_2 = 1$ A. The current in the center branch is $I_1 - I_2 = 1$ A.

The currents do not have to be restricted to the *windows* in order to result in a valid set of simultaneous equations, although that is the usual case with the mesh current method. For example, see Figure 2-8. These are valid *loop* currents.

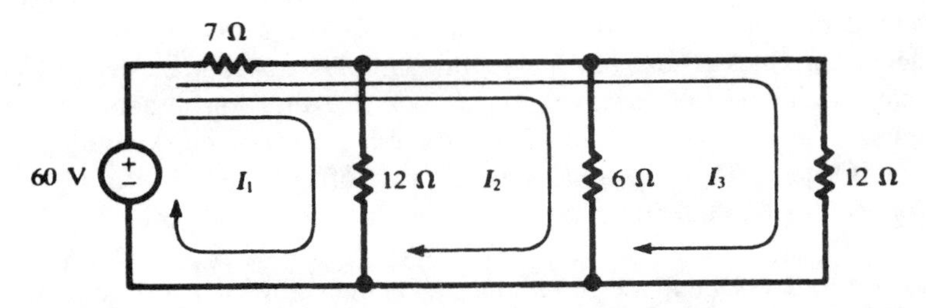

Figure 2-8 Alternate example for valid loop currents

Remember

The applicable rule is that each element in the network must have a current or a combination of currents and no two elements in different branches can be assigned the same current or the same combination of currents.

The simultaneous equations from Example 2.9 can be written in matrix form.

$$\begin{bmatrix} 15 & -10 \\ -10 & 12 \end{bmatrix}\begin{bmatrix} I_1 \\ I_2 \end{bmatrix} = \begin{bmatrix} 20 \\ -8 \end{bmatrix}$$

Once in matrix form, linear algebra methods may be used to solve the mesh equations.

Node Voltage Method

The network shown in Figure 2-9 contains 5 nodes, where *4* and *5* are simple nodes and *1*, *2*, and *3* are principal nodes. In the node voltage method, one of the principal nodes is selected as the reference and equations based on KCL are written at the other principal nodes. At each of

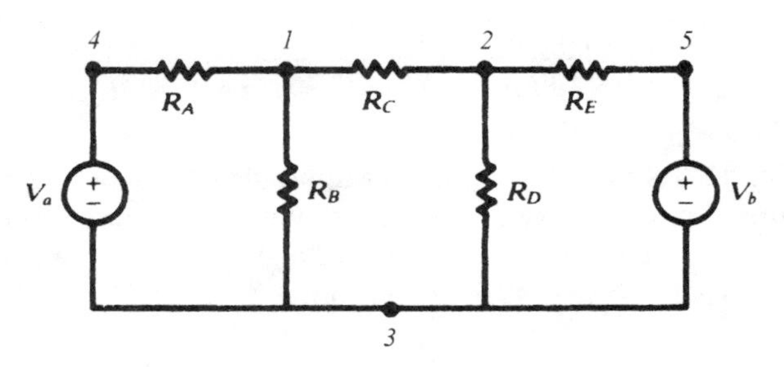

Figure 2-9 Node voltage example

these other principal nodes, a voltage is assigned where it is understood that this is a voltage with *respect to the reference node*. These *node voltages* are the unknowns and, when determined by a suitable method, result in the network solution.

The network is redrawn in Figure 2-10.

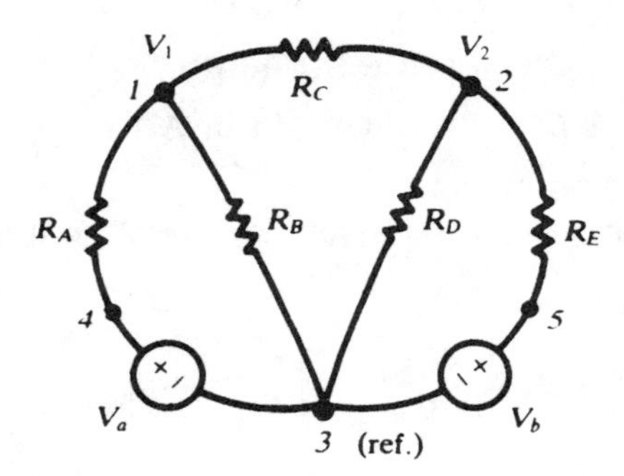

Figure 2-10 Redrawn circuit from Figure 2-9

Node *3* is selected as the reference for voltages V_1 and V_2. KCL requires that the total current out of node *1* be zero:

$$\frac{V_1 - V_a}{R_A} + \frac{V_1}{R_B} + \frac{V_1 - V_2}{R_C} = 0$$

Similarly, the total current out of node *2* must be zero:

$$\frac{V_2 - V_1}{R_C} + \frac{V_2}{R_D} + \frac{V_2 - V_b}{R_E} = 0$$

(Applying KCL in this form does not imply that the actual branch currents all are directed out of either node. Indeed, the current in branch *1–2* is necessarily directed *out of* one node and *into* the other.) Putting the two equations for V_1 and V_2 in matrix form,

$$\begin{bmatrix} \frac{1}{R_A} + \frac{1}{R_B} + \frac{1}{R_C} & -\frac{1}{R_C} \\ -\frac{1}{R_C} & \frac{1}{R_C} + \frac{1}{R_D} + \frac{1}{R_E} \end{bmatrix} \begin{bmatrix} V_1 \\ V_2 \end{bmatrix} = \begin{bmatrix} \frac{V_a}{R_A} \\ \frac{V_b}{R_E} \end{bmatrix}$$

Note the symmetry of the coefficient matrix. The 1,1-element contains the sum of the reciprocals of all resistances connected to node *1*; the 2,2-element contains the sum of the reciprocals of all resistances connected to node *2*. The 1,2- and 2,1-elements are each equal to the *negative* of the sum of the reciprocals of the resistances of all branches joining nodes *1* and node *2* (there is just one such branch in the present circuit). On the right-hand side, the current matrix contains V_a/R_A and V_b/R_E, the driving currents. Both these terms are taken as positive because they both drive a current *into* a node.

Example 2.10 Solve the circuit of Example 2.9 using the node voltage method.

Solution: The circuit is redrawn in Figure 2-11. With two principal nodes, only one equation is required. Assuming the currents are all directed out of the upper node and the bottom node is the reference,

$$\frac{V_1 - 20}{5} + \frac{V_1}{10} + \frac{V_1 - 8}{2} = 0$$

Solving, $V_1 = 10$ V. Then $I_1 = (10 - 20)/5 = -2$ A (the negative sign indicates that the current I_1 flows into node *1*); $I_2 = (10 - 8)/2 = 1$ A; $I_3 = 10/10 = 1$ A.

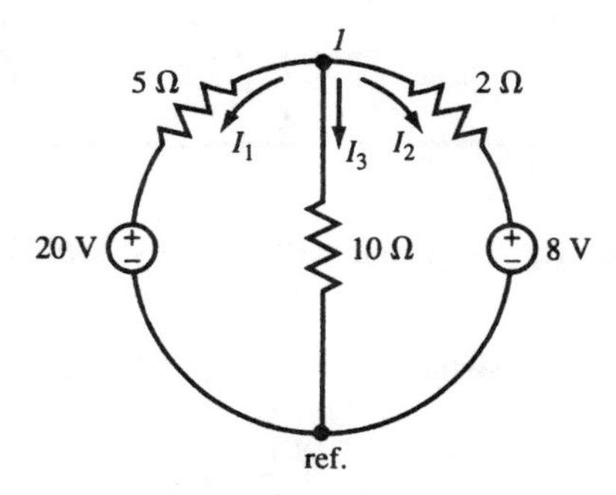

Figure 2-11 Redrawn circuit from Figure 2-7

Network Reduction

The mesh current and node voltage methods are the principal techniques of circuit analysis. However, the equivalent resistance of series and parallel branches, combined with the voltage and current division rules, provide another method of analyzing a network. This method is tedious and usually requires the drawing of several additional circuits. Even so, the process of reducing the network provides a very clear picture of the overall functioning of the network in terms of voltages, currents, and power. The reduction begins with a scan of the network to pick out series and parallel combinations of resistors.

Example 2.11 Obtain the total power supplied by the 60-V source and the power absorbed in each resistor in the network of Figure 2-12.

Solution: Begin by picking out the series and parallel circuits and calculating the equivalent series and parallel resistances.

$$R_{ab} = 7 + 5 = 12\ \Omega$$

$$R_{cd} = \frac{(12)(6)}{12 + 6} = 4\ \Omega$$

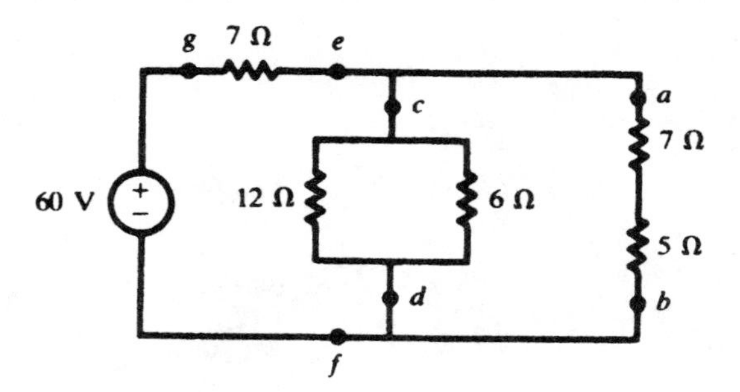

Figure 2-12 Network reduction example circuit

These two equivalent resistances are in parallel (Figure 2-13) giving

$$R_{ef} = \frac{(12)(4)}{12 + 4} = 3\ \Omega$$

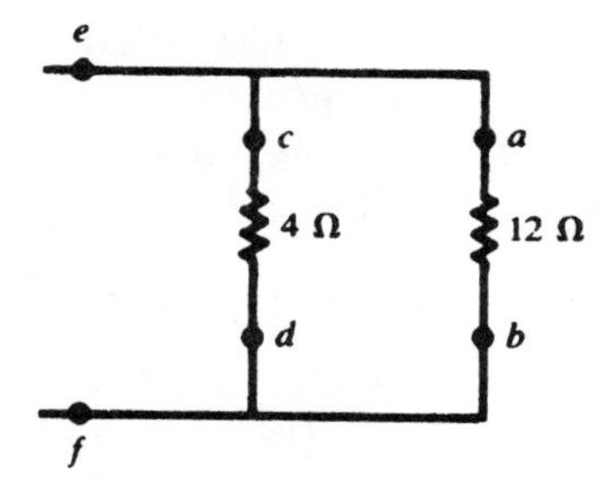

Figure 2-13 Redrawn equivalent circuit

Then this 3-Ω equivalent resistance is in series with the 7-Ω resistor (Figure 2-14), so that for the entire circuit,

$$R_{eq} = 7 + 3 = 10\ \Omega$$

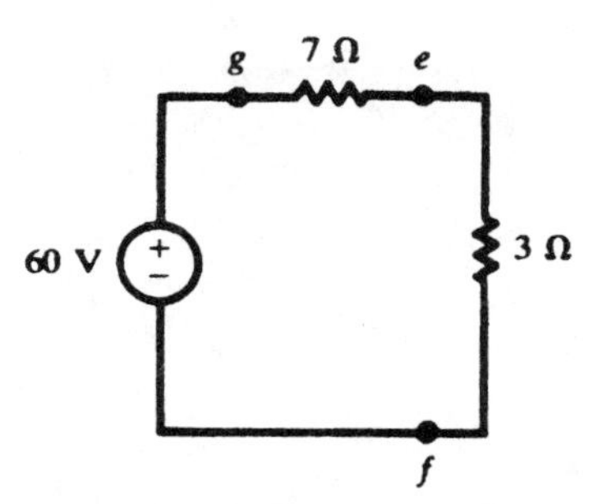

Figure 2-14 Final network reduction

The total power absorbed, which equals the total power supplied by the source, can now be calculated as

$$P_T = \frac{V^2}{R_{eq}} = \frac{(60)^2}{10} = 360\ \text{W}$$

This power is divided between R_{ge} and R_{ef} as follows:

$$P_{ge} = P_{7\Omega} = \frac{7}{7+3}(360) = 252 \text{ W}$$

$$P_{ef} = \frac{3}{7+3}(360) = 108 \text{ W}$$

Power P_{ef} is further divided between R_{cd} and R_{ab} as follows:

$$P_{cd} = \frac{12}{4+12}(108) = 81 \text{ W}$$

$$P_{ab} = \frac{4}{4+12}(108) = 27 \text{ W}$$

Finally, these powers are divided between the individual resistances as follows:

$$P_{12\Omega} = \frac{6}{12+6}(81) = 27 \text{ W}$$

$$P_{6\Omega} = \frac{12}{12+6}(81) = 54 \text{ W}$$

$$P_{7\Omega} = \frac{7}{7+5}(27) = 15.75 \text{ W}$$

$$P_{7\Omega} = \frac{5}{7+5}(27) = 11.25 \text{ W}$$

Superposition

A linear network that contains two or more independent sources can be analyzed to obtain various voltages and branch currents by allowing the sources to act one at a time, then superposing the results. This principle applies because of the linear relationship between current and voltage. With *dependent* sources, superposition can be used only when the control functions are external to the network containing the sources, so that the controls are unchanged as the sources act one at a time.

Voltage sources to be suppressed while a single source acts are replaced by short circuits; open circuits replace current sources.

Superposition cannot be directly applied to the computation of power because power in an element is proportional to the square of the current or the square of the voltage, which is nonlinear.

Example 2.12 Compute the current in the 23-Ω resistor of Figure 2-15.

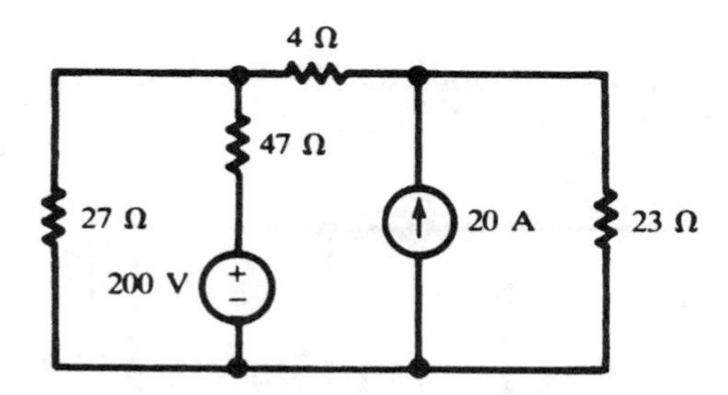

Figure 2-15 Superposition example circuit

Solution: With the 200-V source acting alone, the 20-A current source is replaced by an open circuit as is shown in Figure 2-16.

$$R_{eq} = 47 + \frac{(27)(4+23)}{27+(4+23)} = 60.5\ \Omega$$

$$I_T = \frac{200}{60.5} = 3.31\ \text{A}$$

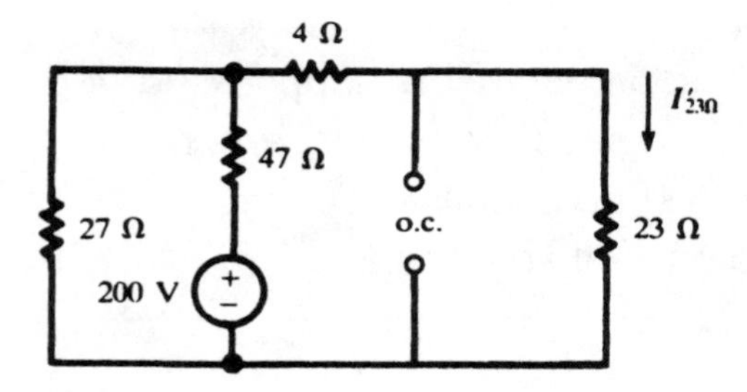

Figure 2-16 Current source is open circuited

The 200-V source's contribution to the current in the 23-Ω resistor is

$$I'_{23\Omega} = \left(\frac{27}{54}\right)(3.31) = 1.65 \text{ A}$$

When the 20-A source acts alone, the 200-V source is replaced by a short circuit as is shown in Figure 2-17.

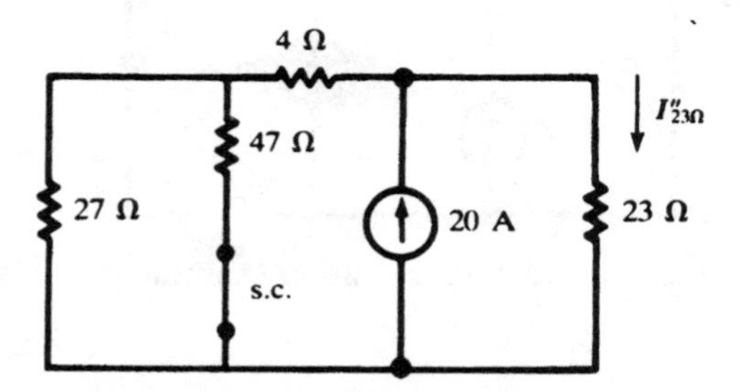

Figure 2-17 Voltage source is short-circuited.

The equivalent resistance to the left of the source is

$$R_{eq} = 4 + \frac{(27)(47)}{27 + 47} = 21.25\ \Omega$$

The 20-A source's contribution to the current in the 23-Ω resistor is

$$I''_{23\Omega} = \left(\frac{21.15}{21.15 + 23}\right)(20) = 9.58 \text{ A}$$

The total current in the 23-Ω resistor is then given by

$$I_{23\Omega} = I'_{23\Omega} + I''_{23\Omega} = 11.23 \text{ A}$$

Thévenin's and Norton's Theorems

A linear, active, resistive network, which contains one or more voltage or current sources, can be replaced by a single voltage source and a series resistance (*Thévenin's theorem*), or by a single current source and a parallel resistance (*Norton's theorem*).

Remember

The voltage is called the *Thévenin equivalent voltage, V′*, and the current is called the *Norton's equivalent current, I′*.

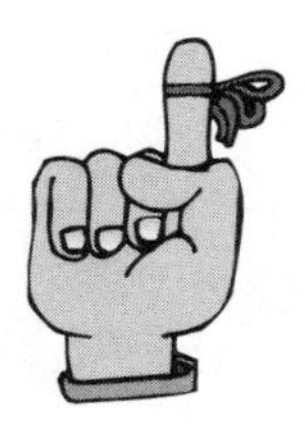

The two resistances are the same, R'. When terminals *ab* in Figure 2-18 are *open-circuited*, a voltage will appear between them.

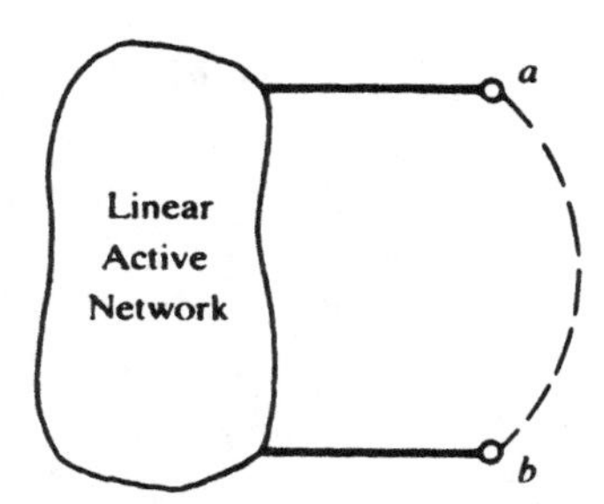

Figure 2-18 Representative linear network

From Figure 2-19, it is evident that this must be the voltage V' of the Thévenin equivalent circuit. If a *short circuit* is applied, a current will result.

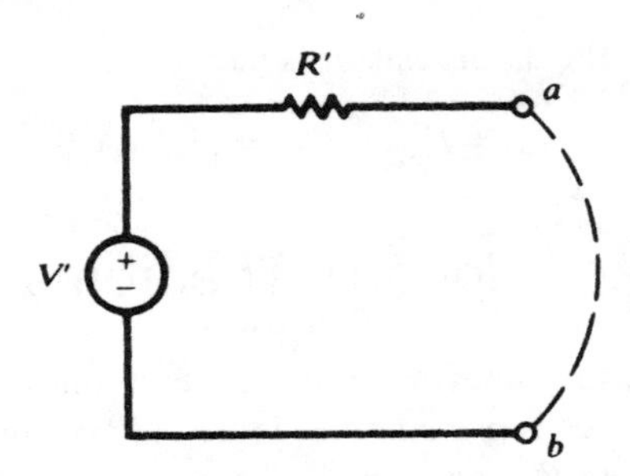

Figure 2-19 Thévenin equivalent circuit

From Figure 2-20, it is evident that this current must be I' of the Norton equivalent circuit.

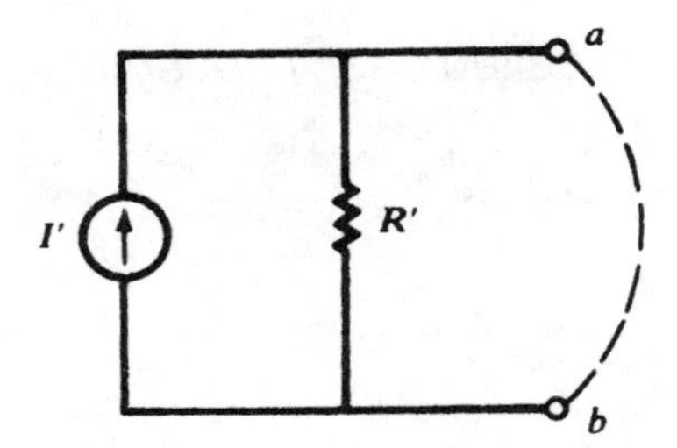

Figure 2-20 Norton equivalent circuit

Now, if the circuits in Figures 2-19 and 2-20 are equivalents of the same active network, they are equivalent to each other. It follows that $I' = V'/R'$. If both V' and I' have been determined from the active network, then $R' = V'/I'$.

Example 2.13 Obtain the Thévenin and Norton equivalent circuits for the active network in Figure 2-21.

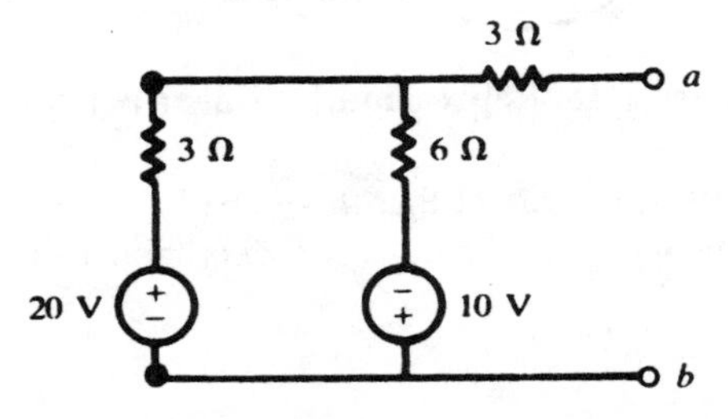

Figure 2-21 Example active linear network

Solution: With terminals *ab* open, the two sources drive a clockwise current through the 3-Ω and 6-Ω resistors (see Figure 2-22).

$$I = \frac{20+10}{3+6} = \frac{30}{9}\ \text{A}$$

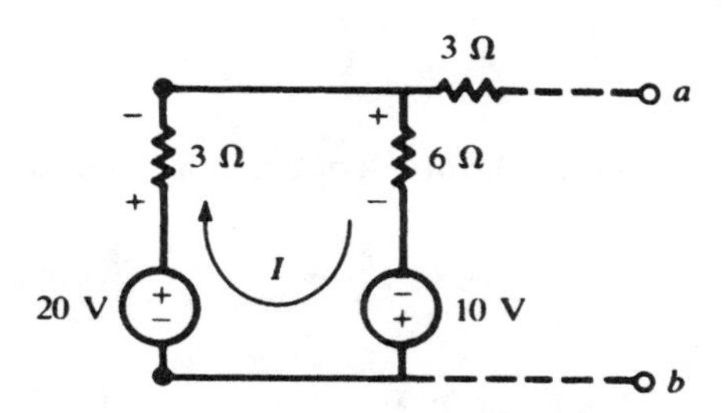

Figure 2-22 Clockwise current flow with open circuited *ab*

Since no current passes through the upper right 3-Ω resistor, the Thévenin voltage can be taken from either active branch:

$$V_{ab} = V' = 20 - \left(\frac{30}{9}\right)3 = 10\ \text{V}$$

or

$$V_{ab} = V' = \left(\frac{30}{9}\right)6 - 10 = 10\ \text{V}$$

The resistance R' is obtained by shorting out the voltage sources (Figure 2-23) and finding the equivalent resistance of this network at terminals *ab*:

$$R' = 3 + \frac{(3)(6)}{9} = 5\ \Omega$$

When a short circuit is applied to the terminals, current I_{sc} results from the two sources. Assuming that it runs through the short from *a* to *b*, we have, by superposition,

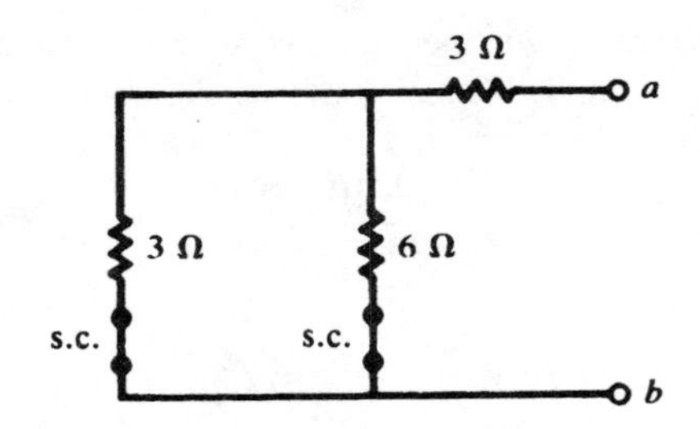

Figure 2-23 Equivalent resistance circuit

$$I_{sc} = I' = \left(\frac{6}{6+3}\right)\left[\frac{20}{3+\dfrac{(3)(6)}{9}}\right] - \left(\frac{3}{3+3}\right)\left[\frac{10}{6+\dfrac{(3)(3)}{6}}\right] = 2\text{ A}$$

Figure 2-24 shows the two equivalent circuits. In the present case, V', R', and I' were obtained independently. Since they are related by Ohm's law, any two may be used to obtain the third.

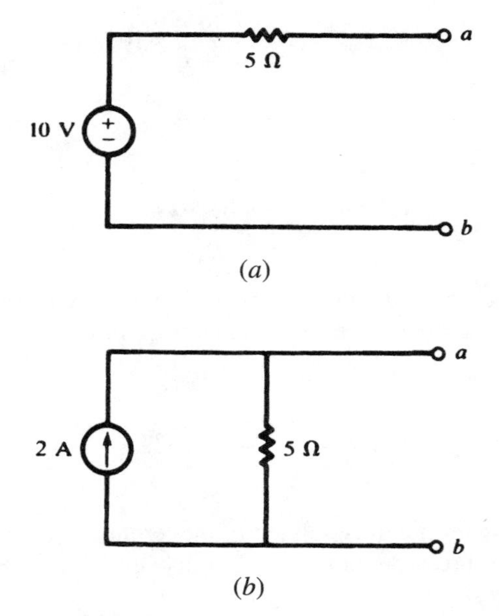

Figure 2-24 (*a*) Thévenin equivalent; (*b*) Norton equivalent

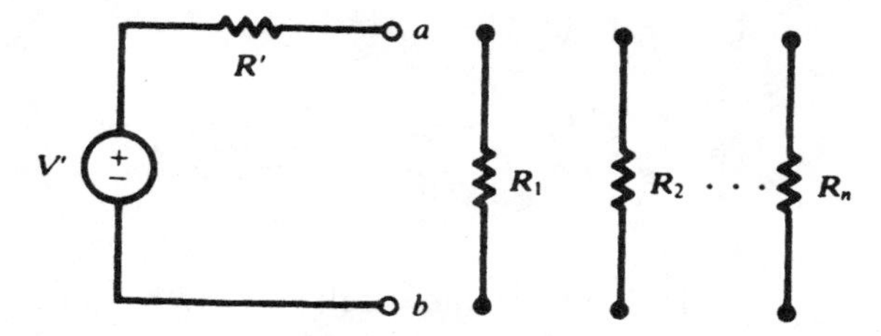

Figure 2-25 Equivalent circuit with varying load conditions

The usefulness of Thévenin and Norton equivalent circuits is clear when an active network is to be examined under a number of load conditions, each represented by a resistor. This is suggested in Figure 2-25, where it is evident that the resistors can be connected one at a time. The resulting current and power are readily obtained. If this were attempted in the original circuit using, for example, network reduction, the task could be very tedious and time consuming.

Maximum Power Transfer

At times, it is desired to obtain the maximum power transfer from an active network to an external load resistor R_L. Assuming that the network is linear, it can be reduced to an equivalent circuit as in Figure 2-26.

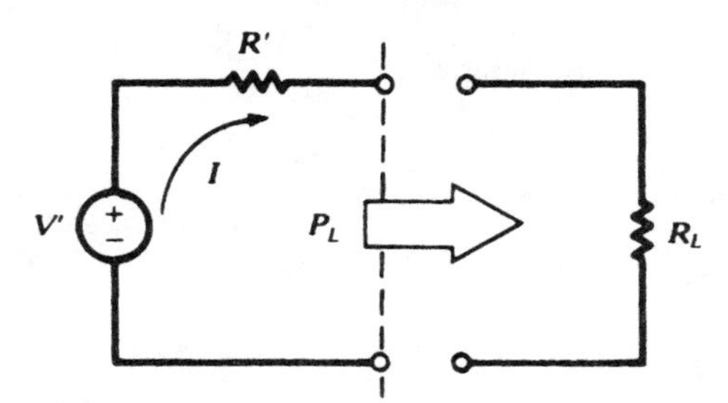

Figure 2-26 Maximum power transfer equivalent circuit

Then,

$$I = \frac{V'}{R' + R_L}$$

The power absorbed by the load is

$$P_L = \frac{(V')^2 R_L}{(R' + R_L)^2} = \frac{(V')^2}{4R'}\left[1 - \left(\frac{R' - R_L}{R' + R_L}\right)^2\right]$$

It is seen that P_L attains its maximum value $(V')^2 / 4R'$, when $R' = R_L$, in which case the power in R' is also $(V')^2/4R'$. Consequently, when the power transferred is a maximum, the efficiency is 50 percent.

It is noted that the condition for maximum power transfer to a load is not the same as the condition for maximum power delivered by the source. The latter happens when $R_L = 0$, in which case power delivered to the load is zero (i.e., at a minimum).

Example 2.13 What value of load resistor R_L connected to terminals *ab* in Figure 2-21 would result in maximum power transfer? What is that maximum transferred power?

Solution: From Figure 2-24(*a*), for maximum power transfer,

$$R_L = R' = 5\ \Omega$$

The maximum power transferred will be

$$P_{L,\max} = (V')^2 / 4R' = \frac{(10)^2}{4(5)} = 5\ \text{W}$$

Important Things to Remember

- ✔ For any closed path in a network, *Kirchhoff's voltage law (KVL)* states that the algebraic sum of the voltages is zero.
- ✔ KCL states that the algebraic sum of the currents entering a node is equal to the sum of the currents leaving that node and is based in conservation of charge.
- ✔ For series connected resistors, the equivalent series resistance is the sum of the resistances; similarly for inductances in series and capacitances in parallel.
- ✔ For parallel connected resistors, the inverse sum is used to find the equivalent parallel resistance; similarly for inductances in parallel and capacitances in series.
- ✔ The mesh current method and node voltage method can result in simultaneous equations that may be solved using linear algebra.
- ✔ Network reduction can be used to simplify a network for more straightforward analysis.
- ✔ In using superposition, inactive voltage sources are replaced by short circuits and open circuits replace inactive current sources.
- ✔ For linear networks, Thévenin and Norton equivalent circuits can be used to simplify calculations when multiple load studies are performed. They can also be used to evaluate the maximum power transfer.

Chapter 3
OPERATIONAL AMPLIFIER CIRCUITS

IN THIS CHAPTER:

- ✔ *Amplifier Models and Feedback*
- ✔ *Operational Amplifiers*
- ✔ *Inverting and Noninverting Circuits*
- ✔ *Voltage Follower*
- ✔ *Differentiator and Integrator Circuits*

Amplifier Models and Feedback

An *amplifier* is a device that magnifies signals. The heart of an amplifier is a source controlled by an input signal. A simplified model of a voltage amplifier is shown in Figure 3-1(a). The input and output reference terminals are often connected together and form a common reference node.

When the output terminal is open we have $v_2 = kv_1$, where k, the multiplying factor, is called the *open circuit gain*.

Resistors R_i and R_o are the input and output resistances of the amplifier, respectively. For a better operation, it is desired that R_i be high and

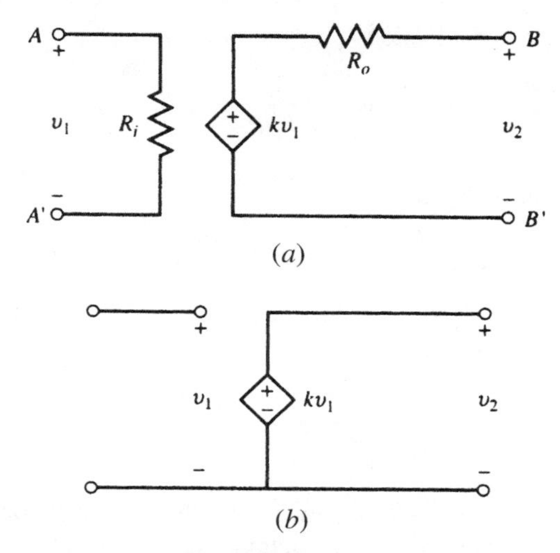

Figure 3-1 Circuit model of a voltage amplifier

R_o be low. In an ideal amplifier, $R_i = \infty$ and $R_o = 0$ as in Figure 3-1(*b*). Deviations from the above conditions can reduce the overall gain.

Example 3.1 In Figure 3-2, a practical voltage source v_s with internal resistance R_s feeds a load R_l through an amplifier with input and output resistances R_i and R_o, respectively. Find the ratio v_2/v_s.

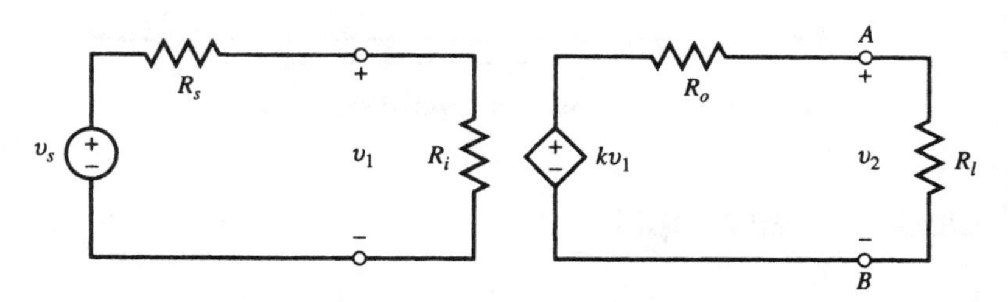

Figure 3-2 Example amplifier circuit

Solution: By voltage division,

$$v_1 = \frac{R_i}{R_i + R_s} v_s$$

Similarly, the output voltage is

$$v_2 = kv_1 \frac{R_l}{R_l + R_o} = k \frac{R_i R_l}{(R_i + R_s)(R_l + R_o)} v_s$$

The ratio is then

$$\frac{v_2}{v_s} = \frac{R_i}{R_i + R_s} \times \frac{R_l}{R_l + R_o} k$$

Note that the open-loop gain is reduced by the factors $R_i/(R_i + R_s)$ and $R_i/(R_i + R_o)$, which also makes the output voltage dependent on the load.

The gain of an amplifier may be controlled by feeding back a portion of its output to its input as is done for the ideal amplifier in Figure 3-3 through the feedback resistor R_2. The feedback resistor affects the overall gain and makes the amplifier less sensitive to variations in k.

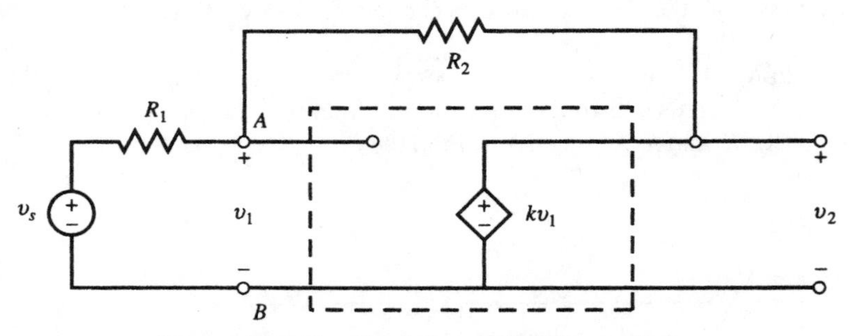

Figure 3-3 Feedback is applied through R_2

Operational Amplifiers

The *operational amplifier* or *op amp* (Figure 3-4) is a device with two input terminals, labeled "+" and "−." These are referred to as the *noninverting* and *inverting* terminals, respectively. The device is also connected to dc power supplies ($+V_{cc}$ and $-V_{cc}$).

$+V_{cc}$ (positive supply)

(inverting input) v^-

v_o (output)

(noninverting input) v^+

$-V_{cc}$ (negative supply)

(common ground)

Figure 3-4 Circuit symbol for an operational amplifier

The common reference for inputs, output, and power supplies resides outside the op amp and is called the *ground.*

The output voltage v_o depends on the differential input voltage $v_d = v^+ - v^-$. Neglecting the capacitive effects, the transfer function is shown in Figure 3-5. In the linear range $v_o = Av_d$ where the open loop gain A is generally very high. V_o saturates at the extremes of $+V_{cc}$ and $-V_{cc}$ when input v_d exceeds the linear range $|v_d| > V_{cc}/A$.

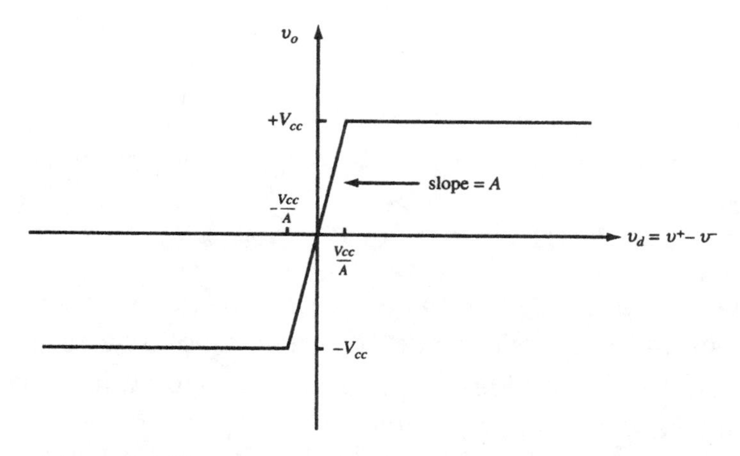

Figure 3-5 Transfer function for an operational amplifier

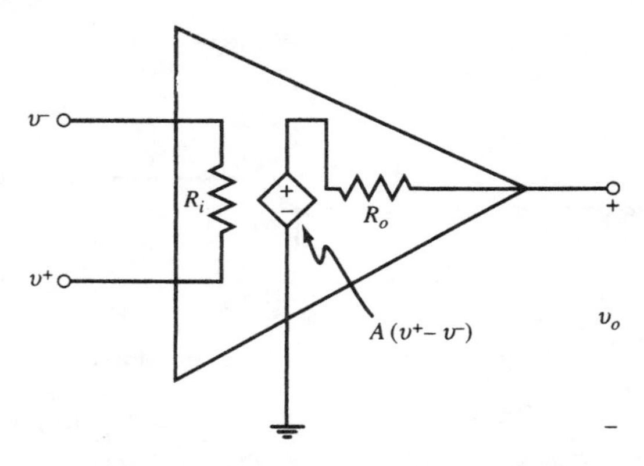

Figure 3-6 Functional circuit model of an op amp

Figure 3-6 shows the model of an op amp in the linear range with power supply connections omitted for simplicity. In practice, R_i is large, R_o is small, and A ranges from 10^5 to several millions. The model of Figure 3-6 is valid so long as the output remains between $+V_{cc}$ and $-V_{cc}$. V_{cc} is generally from 5 to 18 volts.

Example 3.2 In the op amp of Figure 3-6, $V_{cc} = 15$ V, $A = 10^5$, and $v^- = 0$. Find the upper limit on the magnitude of v^+ for linear operation.

Solution: For linear operation,

$$|v_o| = |10^5 v^+| < 15 \text{ V}$$

Therefore,

$$|v^+| < 15 \times 10^{-5} \text{ V} = 150\ \mu\text{V}$$

You Need to Know 

In an *ideal* operational amplifier, R_i and A are assumed to be infinite and R_o is assumed to be zero. Therefore, the ideal op amp draws zero current at its inverting and noninverting inputs.

In most practical amplifying circuits, these assumptions simplify analysis greatly and lead to negligible error. Also, when feedback is used to prevent saturation, these inputs are effectively at the same voltage. Throughout the remaining part of this chapter, we assume op amps are ideal and analyze amplifier circuits that operate in the linear range.

Inverting and Noninverting Circuits

In an *inverting circuit*, the input signal is connected through R_1 to the inverting terminal of the op amp and the output terminal is connected back through a feedback resistor R_2 to the inverting terminal (Figure 3-7). The noninverting terminal A of the op amp is grounded (zero volts). With feedback, the inverting and noninverting inputs are at the same voltage so B is also effectively grounded. Also, for an ideal op amp, the current entering node B is zero.

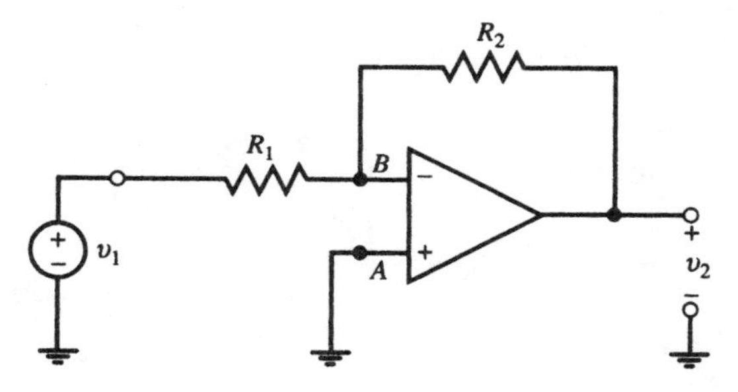

Figure 3-7 Inverting amplifier circuit

To find the gain of the inverting circuit v_2/v_1, apply KCL to the currents arriving at node B:

$$\frac{v_1}{R_1}+\frac{v_2}{R_2}=0 \quad\Rightarrow\quad \frac{v_2}{v_1}=-\frac{R_2}{R_1}$$

The gain is negative and is determined by the choice of resistors only. The input resistance of the circuit is R_1. For an ideal op amp, the output resistance is zero.

Example 3.3 For the circuit in Figure 3-7, if $v_1 = 1.5$ V, $R_1 = 10$ kΩ, and the feedback resistor $R_2 = 40$ kΩ, what is the output voltage v_2?

Solution: The gain of the inverting circuit is

$$\frac{v_2}{v_1} = -\frac{R_2}{R_1} = -\frac{40\ \text{k}\Omega}{10\ \text{k}\Omega} = -4$$

Therefore, the output voltage is

$$v_2 = -\frac{R_2}{R_1} v_1 = -4(1.5) = -6\ \text{V}$$

Note that for this to be in the linear region of the operational amplifier, $V_{cc} > 6$ V.

In a *noninverting circuit*, the input signal arrives at the noninverting terminal of the op amp. The inverting terminal is connected to the output through R_2 and also to the ground through R_1 (see Figure 3-8).

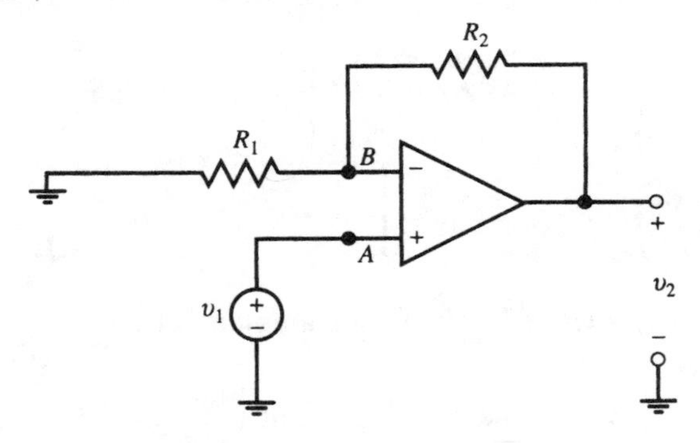

Figure 3-8 Noninverting amplifier circuit

To find the gain v_2/v_1, apply KCL at node B. Note that the terminals A and B are both at v_1 and the op amp draws no current.

$$\frac{v_1}{R_1} + \frac{v_1 - v_2}{R_2} = 0 \quad \Rightarrow \quad \frac{v_2}{v_1} = 1 + \frac{R_2}{R_1}$$

The gain v_2/v_1 is positive and greater than or equal to one. The input resistance of the circuit is infinite, as the op amp draws no current.

Example 3.4 Find v_2/v_1 in the circuit shown in Figure 3-9.

Solution: First, find v_A by dividing v_1 between the 10-kΩ and 5-kΩ resistors.

$$v_A = \frac{5}{5+10}v_1 = \frac{1}{3}v_1$$

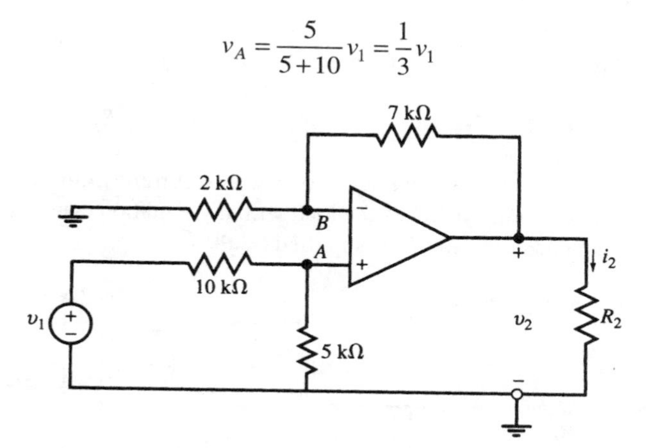

Figure 3-9 Noninverting circuit for Example 3.4

Using the gain expression, we get

$$v_2 = \left(1+\frac{7}{2}\right)v_A = \frac{9}{2}v_A = \left(\frac{9}{2}\right)\left(\frac{1}{3}v_1\right) = 1.5v_1$$

Therefore,

$$v_2 / v_1 = 1.5$$

Voltage Follower

The op amp in the circuit of Figure 3-10 provides a unity gain amplifier in which $v_2 = v_1$ since $v_1 = v^+$, $v_2 = v^-$, and $v^+ = v^-$. The output v_2 follows the input v_1. By supplying i_l to R_l, the op amp eliminates the loading effect of R_l on the voltage source. It therefore functions as a buffer.

Example 3.5 (*a*) Find i_s, v_l, v_2, and i_l in Figure 3-10. (*b*) Compare these results with those obtained when source and load are connected directly as in Figure 3-11.

Solution:

(*a*) With the op amp present (Figure 3-10), we have

$$i_s = 0 \qquad v_1 = v_s \qquad v_2 = v_1 = v_s \qquad i_s = v_s / R_l$$

The voltage follower op amp does not draw any current from the signal source v_s. Therefore, v_s reaches the load with no reduction caused by the load current. The current in R_l is supplied by the op amp.

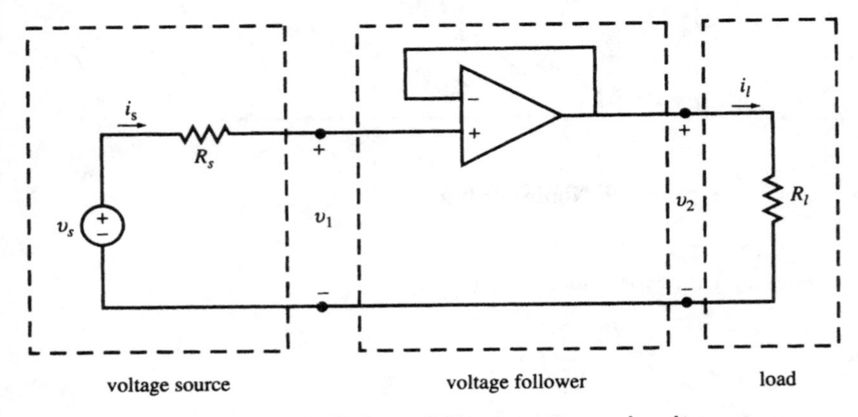

Figure 3-10 Voltage follower op amp circuit

(*b*) With the op amp removed (Figure 3-11), we have

$$i_s = i_l = \frac{v_s}{R_l + R_s} \quad \text{and} \quad v_1 = v_2 = \frac{R_l}{R_l + R_s} v_s$$

The current drawn by R_l goes through R_s and produces a drop in the voltage reaching it. The load voltage v_2 therefore depends on R_l.

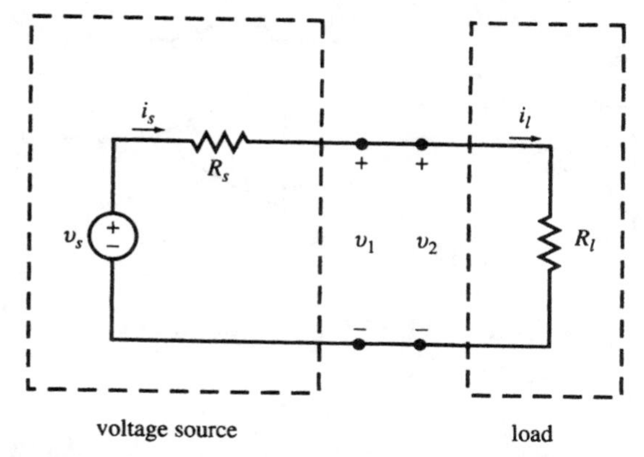

Figure 3-11 Direct connection from Example 3.5

Differentiator and Integrator Circuits

By putting an inductor in place of the feedback resistor for an inverting amplifier, the derivative of the input signal is produced at the output. Figure 3-12 shows the resulting *differentiator circuit.*

To obtain the input-output relationship, apply KCL to currents arriving at the inverting node:

$$\frac{v_1}{R} + \frac{1}{L}\int_{-\infty}^{t} v_2 dt = 0$$

Solving for v_2 (taking the differential with respect to time) gives

$$v_2 = \frac{L}{R}\frac{dv_1}{dt}$$

This shows that the output is a weighted derivative of the input.

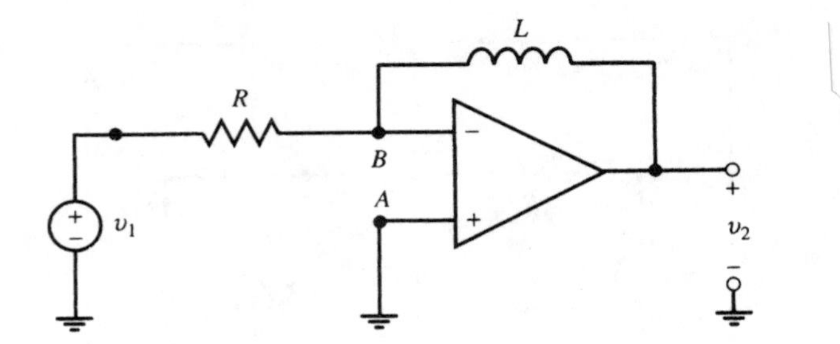

Figure 3-12 Differentiator circuit

By replacing the feedback resistor in the inverting amplifier with a capacitor, the basic *integrator circuit* of Figure 3-13 will result. To obtain the input-output relationship, apply KCL at the inverting node:

$$\frac{v_1}{R} + C\frac{dv_2}{dt} = 0 \quad \text{from which} \quad \frac{dv_2}{dt} = -\frac{1}{RC}v_1$$

and

$$v_2 = -\frac{1}{RC}\int_{-\infty}^{t} v_1 dt$$

In other words, the output is equal to the integral of the input multiplied by a gain factor of $-1/RC$.

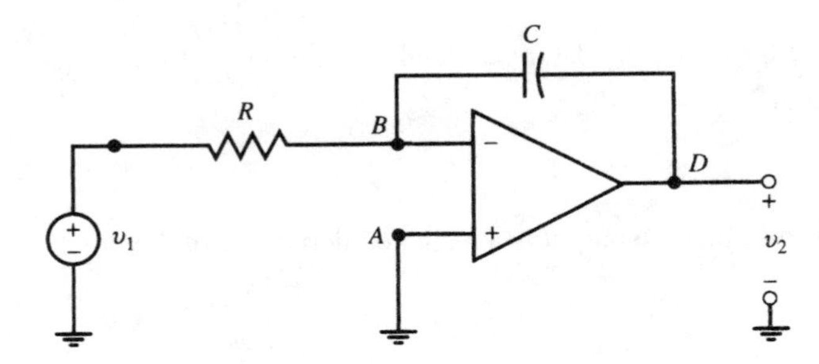

Figure 3-13 Integrator circuit

Important Things to Remember

- ✔ An *amplifier* is a device that magnifies signals.
- ✔ Ideal amplifiers have input resistance $R_i = \infty$ and output resistance $R_o = 0$.
- ✔ Feedback resistors are used to control the overall gain and make the amplifier less sensitive to variations in open-loop gain k.
- ✔ When operational amplifiers are used without feedback, their very high gain (typically greater than 10^5) causes them to saturate at an output voltage of $\pm V_{cc}$.
- ✔ With feedback, the input voltages at the "+" and "–" terminals of the op amp are effectively the same.
- ✔ Evaluation of KCL at the input terminals of the op amp is an often-used method in operational amplifier analysis and takes advantage of the ideal current of the op amp being zero.

Chapter 4
WAVEFORMS AND TRANSIENT FIRST-ORDER CIRCUITS

IN THIS CHAPTER:

- ✔ *Periodic Functions*
- ✔ *Sinusoidal Functions*
- ✔ *Nonperiodic Functions*
- ✔ *Transient First-Order RC Circuits*
- ✔ *Transient First-Order RL Circuits*
- ✔ *Response of a First-Order Circuit to a Pulse*
- ✔ *Response of a First-Order Circuit to an Impulse*

Periodic Functions

A signal $v(t)$ is periodic with *period* T if $v(t) = v(t + T)$ for all time t. Two types of periodic functions, which are specified in Figures 4-1 and 4-2 for one period T and corresponding graphs, are discussed next.

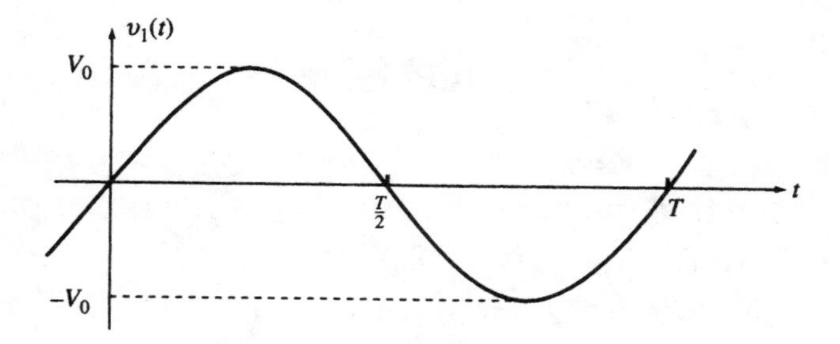

Figure 4-1 Sinusoidal function

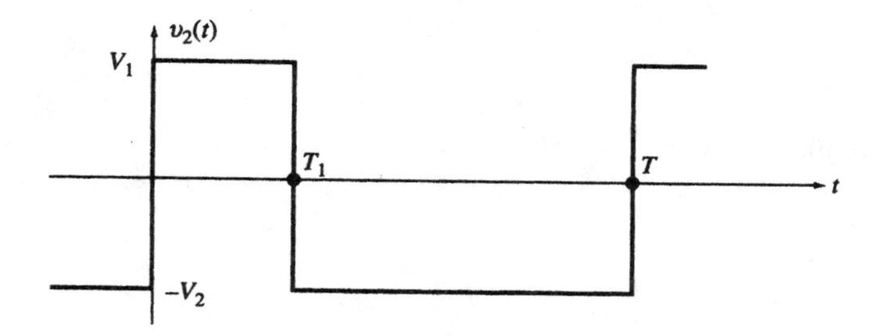

Figure 4-2 Periodic pulse function

Sinusoidal Functions

A sinusoidal voltage is given by

$$v(t) = V_0 \cos(\omega t + \theta)$$

where V_0 is the amplitude, ω is the angular frequency and θ is the phase angle in radians or degrees.

NOTE!

The angular frequency ω has units of radians per second and may be expressed in terms of the period T or the frequency f. The frequency $f \equiv 1/T$ and has units of *hertz*, Hz, or cycles per second. The relationship to angular frequency is $\omega = 2\pi f$.

To summarize the relationships between ω, f and T,

$$\omega = 2\pi/T = 2\pi f \qquad f = 1/T = \omega/2\pi \qquad T = 1/f = 2\pi/\omega$$

Example 4.1 Graph each of the following functions and specify period and frequency: (*a*) $v_1(t) = \cos t$; (*b*) $v_2(t) = \sin t$; (*c*) $v_3(t) = 2\cos 2\pi t$

Solution:

(*a*) See Figure 4-3(*a*). $T = 2\pi = 6.2832$ s and $f = 0.159$ Hz

(*b*) See Figure 4-3(*b*). $T = 2\pi = 6.2832$ s and $f = 0.159$ Hz

(*c*) See Figure 4-3(*c*). $T = 1$s and $f = 1$ Hz

If the function $v(t) = \cos\omega t$ is delayed by τ seconds, we get $v(t - \tau) = \cos\omega(\tau - t) = cos(\omega t - \theta)$, where $\theta = \omega\tau$.

The delay shifts the graph of $v(t)$ to the right by an amount of τ seconds, which corresponds to a *phase lag* of θ.

A *time shift* of τ seconds to the left on the graph produces $v(t + \tau)$, resulting in a leading phase angle called an *advance*. Conversely, a *phase shift* of θ corresponds to a time shift of τ. Therefore, for a given phase shift, the higher the frequency, the smaller is the required time shift.

The sum of two periodic functions with respective periods T_1 and T_2 is a periodic function if a common period $T = n_1T_1 = n_2T_2$, where n_1 and n_2 are integers, can be found. This requires $T_1/T_2 = n_2/n_1$ to be a rational number. Otherwise, the sum is not a periodic function.

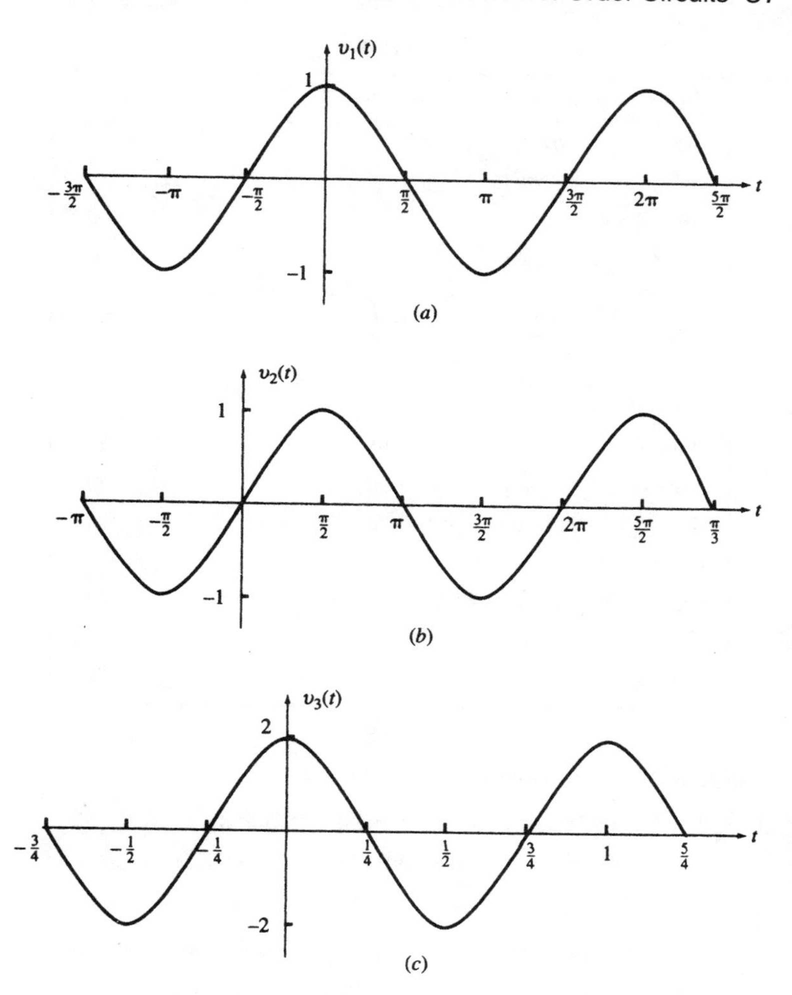

Figure 4-3 Example sinusoids

Trigonometric functions that are useful in sinusoidal circuit analysis are given in Table 4-1.

Example 4.2 Express $v(t) = \cos 5t \sin(3t + 45°)$ as the sum of two cosine functions and find its period.

Table 4-1

$\sin a = -\sin(-a)$	(5a)
$\cos a = \cos(-a)$	(5b)
$\sin a = \cos(a - 90°)$	(5c)
$\cos a = \sin(a + 90°)$	(5d)
$\sin 2a = 2 \sin a \cos a$	(6a)
$\cos 2a = \cos^2 a - \sin^2 a = 2\cos^2 a - 1 = 1 - 2\sin^2 a$	(6b)
$\sin^2 a = \dfrac{1 - \cos 2a}{2}$	(7a)
$\cos^2 a = \dfrac{1 + \cos 2a}{2}$	(7b)
$\sin(a + b) = \sin a \cos b + \cos a \sin b$	(8a)
$\cos(a + b) = \cos a \cos b - \sin a \sin b$	(8b)
$\sin a \sin b = \frac{1}{2}\cos(a - b) - \frac{1}{2}\cos(a + b)$	(9a)
$\sin a \cos b = \frac{1}{2}\sin(a + b) + \frac{1}{2}\sin(a - b)$	(9b)
$\cos a \cos b = \frac{1}{2}\cos(a + b) + \frac{1}{2}\cos(a - b)$	(9c)
$\sin a + \sin b = 2 \sin\frac{1}{2}(a + b) \cos\frac{1}{2}(a - b)$	(10a)
$\cos a + \cos b = 2 \cos\frac{1}{2}(a + b) \cos\frac{1}{2}(a - b)$	(10b)

Solution: Using the equations from Table 4-1,

$$v(t) = \cos 5t \sin(3t + 45°) = [\sin(8t + 45°) - \sin(2t - 45°)]/2$$

$$v(t) = [\cos(8t - 45°) - \cos(2t + 45°)]/2$$

The period of $v(t)$ is π.

A periodic function $f(t)$, with a period T, has an average value F_{avg} given by

$$F_{avg} = \frac{1}{T}\int_0^T f(t)\, dt = \frac{1}{T}\int_{t_0}^{t_0+T} f(t)\, dt$$

The *root-mean-square* (*rms*) or *effective value* of $f(t)$ during the same period is defined by

$$F_{rms} = \left[\frac{1}{T}\int_{t_0}^{t_0+T} f^2(t)\, dt\right]^{1/2}$$

Average and effective values of periodic functions are normally computed over one period.

Example 4.3 Find the average and effective values of the cosine wave $v(t) = V_m \cos(\omega t + \theta)$.

Solution: Using the above equations,

$$V_{avg} = \frac{1}{T}\int_0^T V_m \cos(\omega t + \theta)\, dt = \frac{V_m}{\omega T}[\sin(\omega t + \theta)]_0^T = 0$$

$$V_{eff}^2 = \frac{1}{T}\int_0^T V_m^2 \cos^2(\omega t + \theta)\, dt = \frac{1}{2T}\int_0^T V_m^2 [1 + \cos 2(\omega t + \theta)]\, dt = \frac{V_m^2}{2}$$

from which

$$V_{eff} = \frac{V_m}{\sqrt{2}} = 0.707\, V_m$$

RMS (effective) values are often used in electrical power analysis.

Nonperiodic Functions

A nonperiodic function cannot be specified for all times by simply knowing a finite segment. Several nonperiodic functions are used as models and building blocks for actual signals in analysis and design of circuits. Examples of some of the functions follow.

The dimensionless *unit step function* is defined by

$$u(t) = \begin{cases} 0 & t < 0 \\ 1 & t > 0 \end{cases}$$

The function is graphed in Figure 4-4. Note that the function in undefined at $t = 0$.

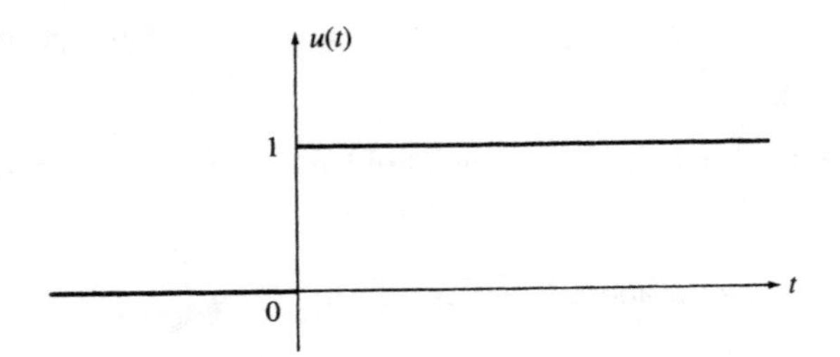

Figure 4-4 Unit step function

To illustrate the use of $u(t)$, assume the switch S in the circuit of Figure 4-5 has been in position 1 for $t < 0$ and is moved to position 2 at $t = 0$. The voltage across terminals A-B may be expressed by $v_{AB} = V_0 u(t)$.

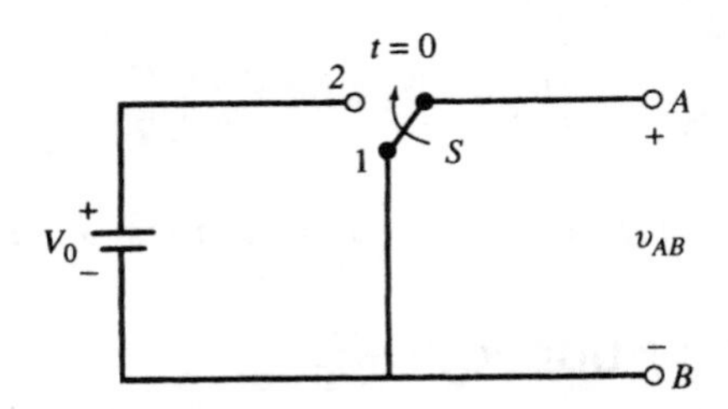

Figure 4-5 Switch causing a unit step voltage at *A*-*B*

Example 4.4 The switch in the circuit of Figure 4-5 is moved to position 2 at $t = t_0$. Express v_{AB} using the unit step function.

Solution: The appearance of V_0 across A-B is delayed until $t = t_0$. Replace the argument t in the step function by $t - t_0$ and so we have $v_{AB} = V_0 u(t - t_0)$

Example 4.5 If the switch in Figure 4-5 is moved to position 2 at $t = 0$ and then moved back to position 1 at $t = 5$ s, express v_{AB} using the unit step function.

Solution: $$v_{AB} = V_0[u(t) - u(t - 5)]$$

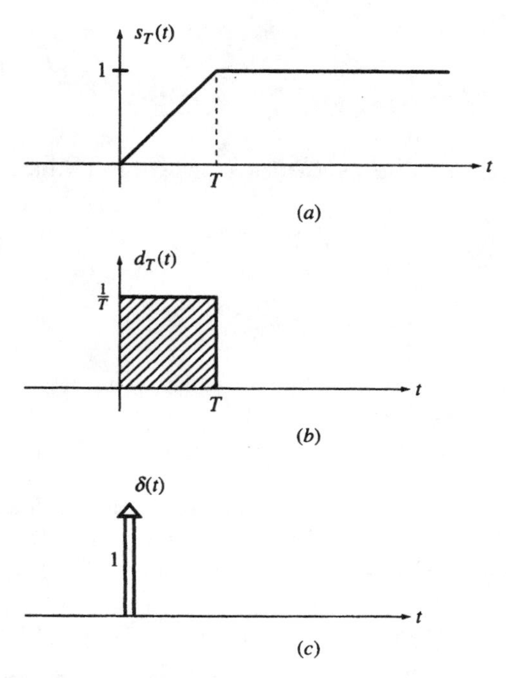

Figure 4-6 Defining the unit impulse function

Consider the function $s_T(t)$ of Figure 4-6(a) which is zero for $t < 0$ and increases uniformly from 0 to 1 in T seconds. Its derivative $d_T(t)$ is a pulse of duration T and height $1/T$, as seen in Figure 4-6(b).

$$d_T(t) = \begin{cases} 0 & t < 0 \\ 1/T & 0 < t < T \\ 0 & t > T \end{cases}$$

If the transition time T is reduced, the pulse in Figure 4-6(b) becomes narrower and taller, but the area under the pulse remains equal to 1. If we let T approach zero, in the limit function $s_T(t)$ becomes a unit step $u(t)$ and its derivative $d_T(t)$ becomes a unit impulse $\delta(t)$ with zero width and infinite height. The unit impulse $\delta(t)$ is shown in Figure 4-6(c).

You Need to Know ✓

The unit impulse or *delta function* is defined by

$$\delta(t) = \begin{cases} 0 & t \neq 0 \\ \infty & t = 0 \\ \int_{-\infty}^{\infty} \delta(t)\, dt = 1 & \end{cases}$$

An impulse, which is the limit of a narrow pulse with an area A, is expressed by $A\delta(t)$. The magnitude A is sometimes called the *strength* of the impulse. A unit impulse which occurs at $t = t_0$ is expressed by $\delta(t - t_0)$.

Another nonperiodic function often found in circuit analysis is the *exponential* $f(t) = e^{at}$ where a is constant (can be complex). The inverse of the constant a has the dimension of time and is called the *time constant* $\tau = 1/a$.

A decaying exponential $e^{-t/\tau}$ is plotted versus t as shown in Figure 4-7.

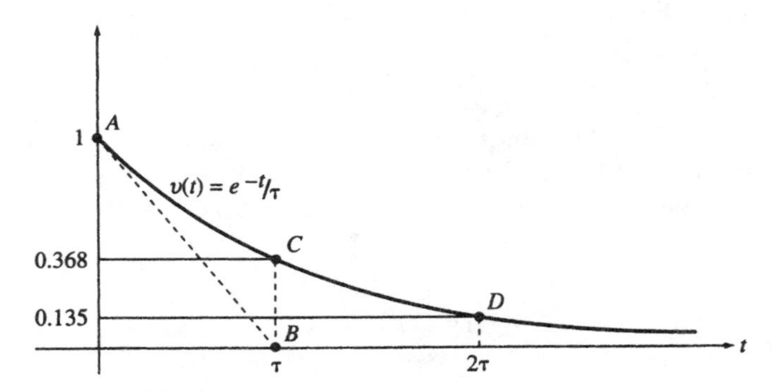

Figure 4-7 Decaying exponential function

The function decays form one at $t = 0$ to zero at $t = \infty$. After τ seconds the function $e^{-t/\tau}$ is reduced to $e^{-1} = 0.368$.

Transient First-Order RC Circuits

Whenever a circuit is switched form one condition to another, either by a change in the applied source or a change in the circuit elements, there is a transitional period during which the branch currents and element voltages change from their former values to new ones. This period is called the *transient*. After the transient has passed, the circuit is said to be in the *steady state*. Now, the linear differential equation that describes the circuit will have two parts to its solution, the *complementary function* (or the *homogeneous solution*) and the *particular solution*. The complementary function corresponds to the transient, and the particular solution to the steady state. The two types of circuits evaluated here are ones involving resistors and capacitors and ones involving resistors and inductors.

Assume a capacitor has a voltage difference V_0 between its plates. When a conducting path R is provided, the stored charge travels through the capacitor from one plate to the other, establishing current i. Thus, the capacitor voltage is v is gradually reduced to zero, at which time the current also becomes zero. In the RC circuit of Figure 4-8(a), $Ri = v$ and $i = -C\, dv/dt$. Eliminating i in both equations gives

$$\frac{dv}{dt} + \frac{1}{RC} v = 0$$

The only function whose linear combination with its derivative can be zero is an exponential function of the form Ae^{st}. Replacing v by Ae^{st} and dv/dt by sAe^{st} gives

$$sAe^{st} + \frac{1}{RC} Ae^{st} = A\left(s + \frac{1}{RC}\right)e^{st} = 0$$

For the equality with zero, this means

$$s = -\frac{1}{RC}$$

Given that $v(0) = A = V_0$, $v(t)$ and $i(t)$ are found to be

$$v(t) = V_0 e^{-t/RC}, t > 0 \qquad (1)$$

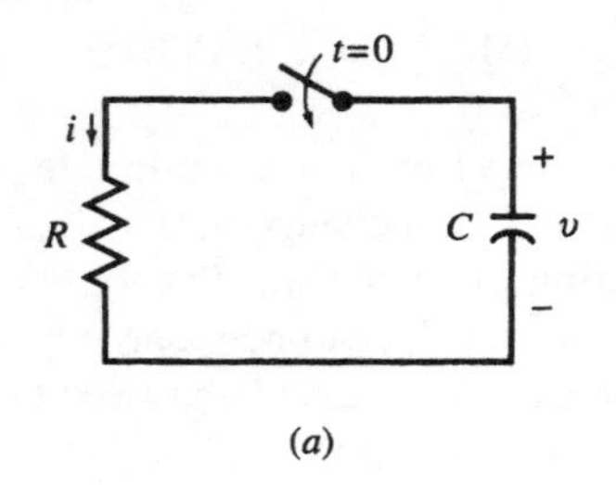

(a)

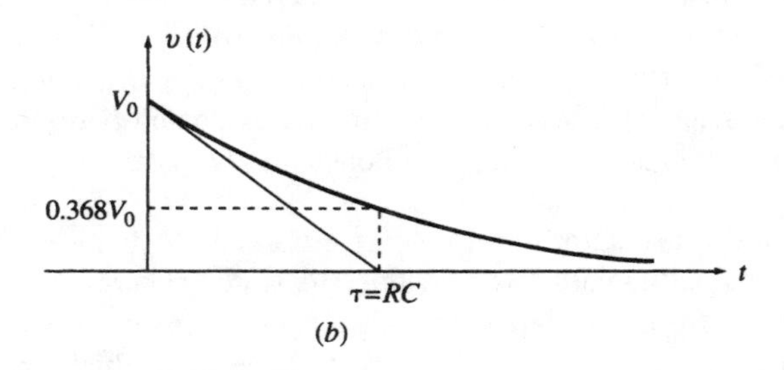

(b)

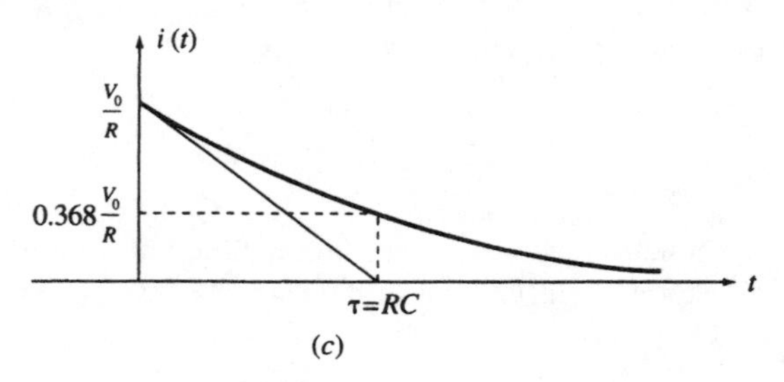

(c)

Figure 4-8 *RC* transient circuit

$$i(t) = -C\frac{dv}{dt} = \frac{V_0}{R}e^{-t/RC}, t > 0 \tag{2}$$

The voltage and current of the capacitor are exponentials with initial values of V_0 and V_0/R, respectively. As time increases, voltage and current decrease to zero with a time constant of $\tau = RC$. See Figures 4-8(*b*) and 4-8(*c*).

Example 4.6 The voltage across a 1-μF capacitor is 10 V for $t < 0$. At $t = 0$, a 1-MΩ resistor is connected across the capacitor terminals (Figure 4-8(a)). Find the time constant τ, the voltage $v(t)$, and its value at $t = 5$ s.

Solution:

$$\tau = RC = 1\text{s}$$
$$v(t) = 10e^{-t}\text{ V}, t > 0$$
$$v(5) = 10e^{-5} = 0.067\text{ V}$$

Example 4.7 A 5-μF capacitor with an initial voltage of 4 V is connected to a parallel combination of a 3-kΩ and a 6-kΩ resistor (Figure 4-9). Find the current in the 6-kΩ resistor.

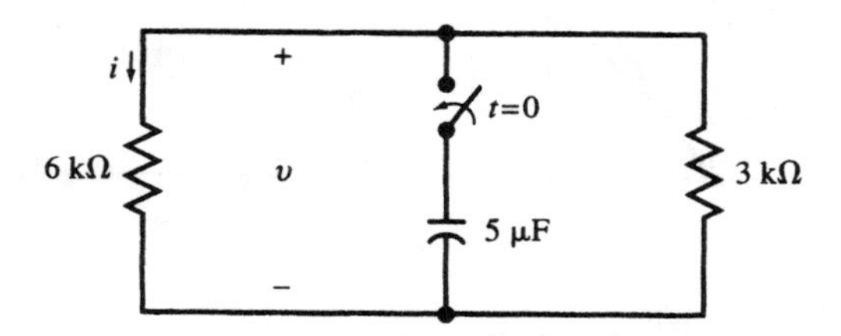

Figure 4-9 Parallel *RC* circuit

Solution: The equivalent resistance of the two parallel resistors is $R = 2$ kΩ. The time constant of the circuit is $RC = 10^{-2}$ s. The voltage and current in the 6-kΩ resistor are, respectively,

$$v(t) = 4e^{-100t}\text{ V}, t > 0$$
$$i(t) = v/6000 = 0.67e^{-100t}\text{ mA}, t > 0$$

Connect an initially uncharged capacitor to a battery with voltage V_0 through a resistor at $t = 0$. The circuit is shown in Figure 4-10(a).

For $t > 0$, KVL around the loop gives $Ri + v = V_0$ which, after substituting $i = C\,dv/dt$, becomes

$$\frac{dv}{dt} + \frac{1}{RC}v = \frac{1}{RC}V_0, t > 0 \tag{3}$$

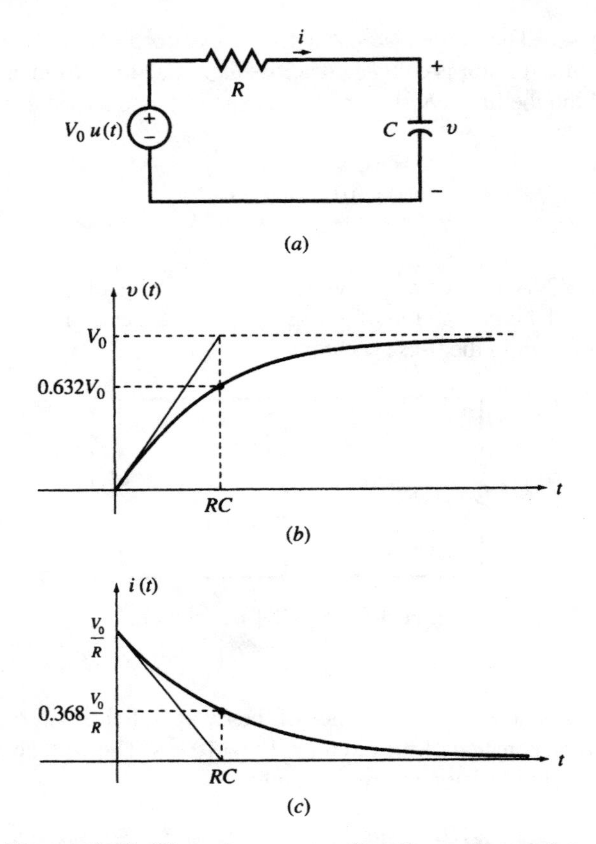

Figure 4-10 Charging a capacitor through a resistor

with the initial condition

$$v(0^+) = v(0^-) = 0 \tag{4}$$

The solution should satisfy both of the above equations. The *particular* solution (or *forced response*) $v_p(t) = V_0$ satisfies (3) but not (4). The *homogeneous* solution (or *natural response*) $v_h(t) = Ae^{-t/RC}$ can be added and its magnitude A adjusted so that the total solution satisfies both (3) and (4):

$$v(t) = v_p(t) + v_h(t) = V_0 + Ae^{-t/RC}$$

From the initial condition, $v(0^+) = V_0 + A = 0 \quad \Rightarrow \quad A = -V_0$. Thus, the total solution is (see Figures 4-10(b) and 4-10(c)).

$$v(t) = V_0(1 - e^{-t/RC})\, u(t) \text{ V} \qquad (5)$$

$$i(t) = \frac{V_0}{R} e^{-t/RC}\, u(t) \text{ A} \qquad (6)$$

Example 4.8 A 4-μF capacitor with in an initial voltage of $v(0^-) = 2V$ is connected to a 12-V battery through a 5-kΩ resistor at $t = 0$. Find the voltage across and the current through the capacitor for $t > 0$.

Solution: The time constant of the circuit is $\tau = RC = 0.02$ s. Following the analysis of Example 4.7, we get

$$v(t) = 12 + Ae^{-50t}$$

From the initial conditions, $v(0^+) = v(0^-) = 12 + A = 2$ or $A = -10$. Thus, for $t > 0$,

$$v(t) = 12 - 10e^{-50t} \text{ V}$$

$$i(t) = \frac{12 - v}{5000} = 2 \times 10^{-3} e^{-50t} \text{ A} = 2e^{-50t} \text{ mA}$$

The current may also be computed from $i = C\, dv/dt$. And so the voltage increases exponentially from an initial value of 2 V to a final value of 12 V, with a time constant of 20 ms, as shown in Figure 4-11(a), while the current decreases from 2 mA to zero as shown in Figure 4-11(b).

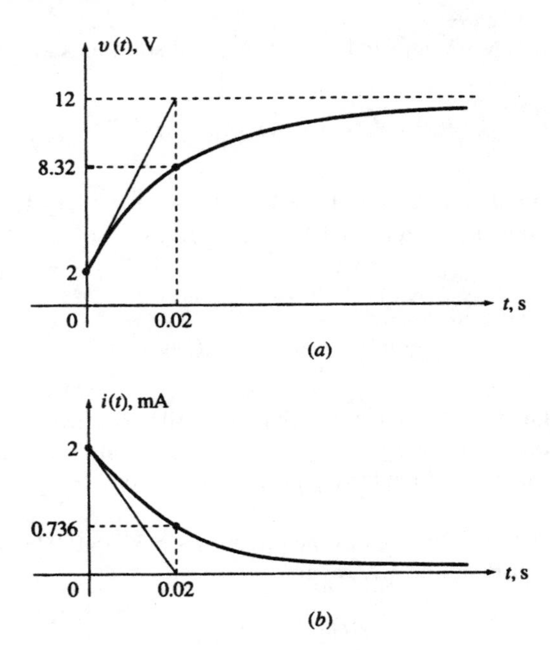

Figure 4-11 ***RC*** **circuit of Example 4.8**

Transient First-Order RL Circuits

In the *RL* circuit of Figure 4-12, assume that at $t = 0$, the current is I_0. For $t > 0$, i should satisfy $Ri + L(di/dt) = 0$, the solution of which is $i = Ae^{st}$. By substitution, we find A and s:

$$A(R + Ls)e^{st} = 0, \quad R + Ls = 0, \quad s = -R/L$$

The initial condition $i(0) = A = I_0$. Then

$$i(t) = I_0 e^{-Rt/L} \text{ for } t > 0$$

The time constant of the circuit is L/R.

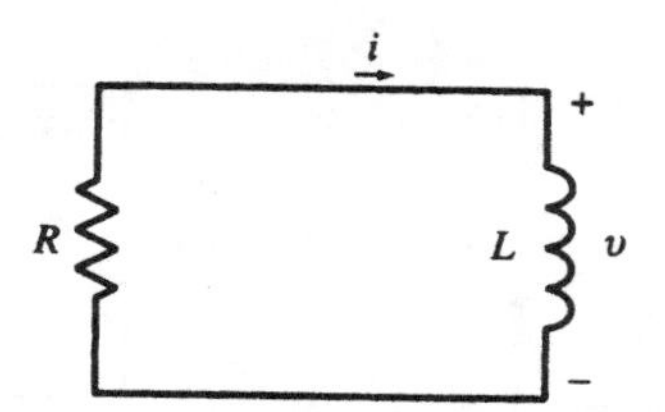

Figure 4-12 ***RL*** **transient circuit**

Example 4.9 The 12-V battery in Figure 4-13 is disconnected at $t = 0$. Find the inductor current and voltage v for all times.

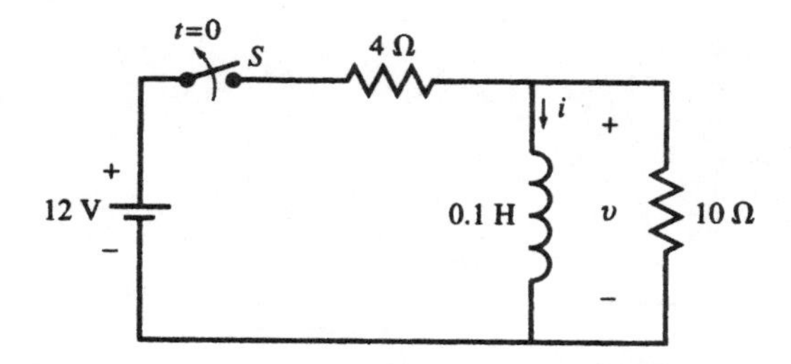

Figure 4-13 ***RL*** **transient circuit example**

Solution: Assume the switch *S* has been closed for a long time. The inductor current is then constant and its voltage is zero. The circuit at $t = 0^-$ is shown in Figure 4-14(a) with $i(0^-) = 12/4 = 3$ A. After the battery is disconnected, for $t > 0$, the circuit will be as shown in Figure 4-14(b). For $t > 0$, the current decreases exponentially from 3 A to zero. The time constant of the circuit is $L/R = 0.01$ s. Using the results of Example 4.8, for $t > 0$, the inductor current and voltage are, respectively,

$$i(t) = 3e^{-100t} \text{ A}, \quad t > 0$$

$$v(t) = L(di / dt) = -30e^{-100t} \text{ V}, \quad t > 0$$

$i(t)$ and $v(t)$ are plotted in Figures 4-15(a) and 4-15(b), respectively.

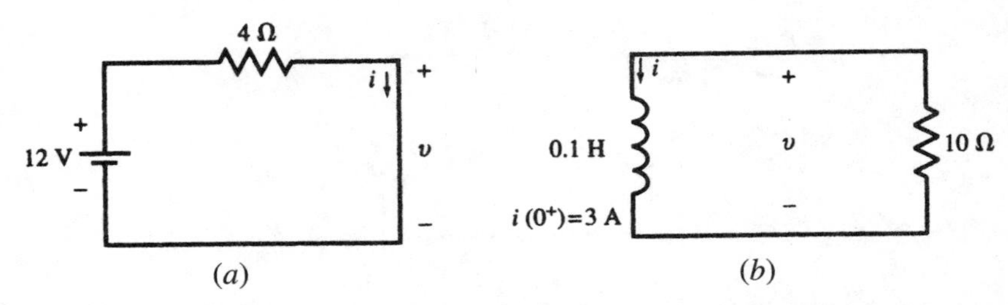

Figure 4-14 Equivalent circuit models before (*a*) and after (*b*) the switch is thrown

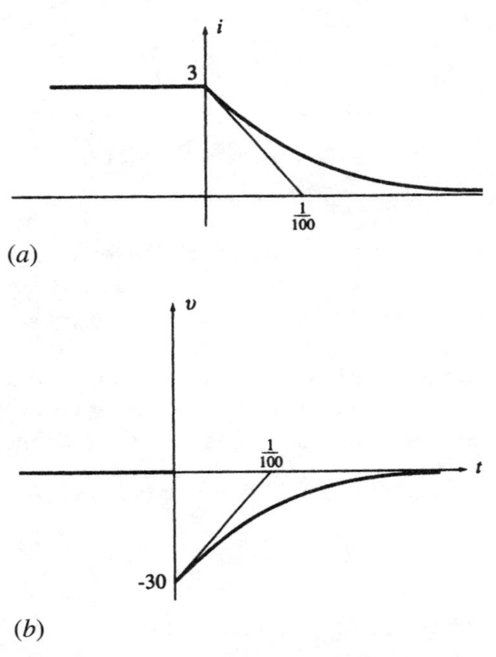

Figure 4-15 *i* and *v* for the transient *RL* circuit

Response of a First-Order Circuit to a Pulse

In this section, we will derive the response of a first-order circuit to a rectangular pulse. The derivation applies to *RC* or *RL* circuits where the input can be a current or a voltage. As an example, we use the series *RC* circuit in Figure 4-16 with the voltage source delivering a pulse of duration T and height V_0. For $t < 0$, v and i are zero. For the duration of the pulse, we use (5) and (6) above:

$$v(t) = V_0(1 - e^{-t/RC}),\ 0 < t < T \tag{7}$$

$$i(t) = \frac{V_0}{R} e^{-t/RC},\ 0 < t < T \tag{8}$$

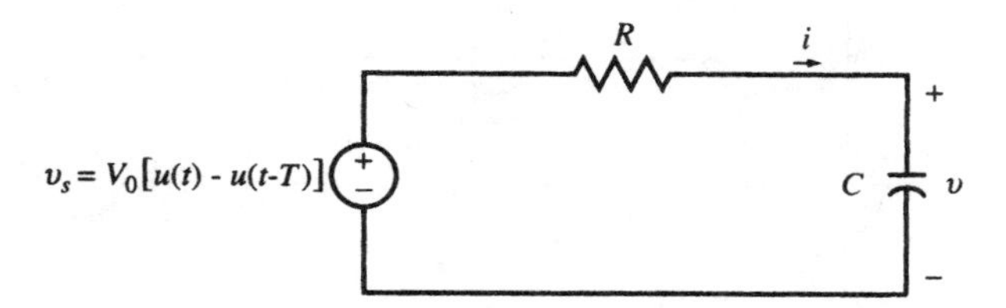

Figure 4-16 *RC* circuit with a pulse source

When the pulse ceases, the circuit is source-free with the capacitor at an initial voltage V_T.

$$V_T = V_0(1 - e^{-T/RC}) \tag{9}$$

Using (1) and (2) above, and taking into account the time shift T,

$$v(t) = V_T e^{-(t-T)/RC},\ t > T \tag{10}$$

$$i(t) = -\frac{V_T}{R} e^{-(t-T)/RC},\ t > T \tag{11}$$

The capacitor voltage and current are plotted in Figures 4-17(a) and 4-17 (b).

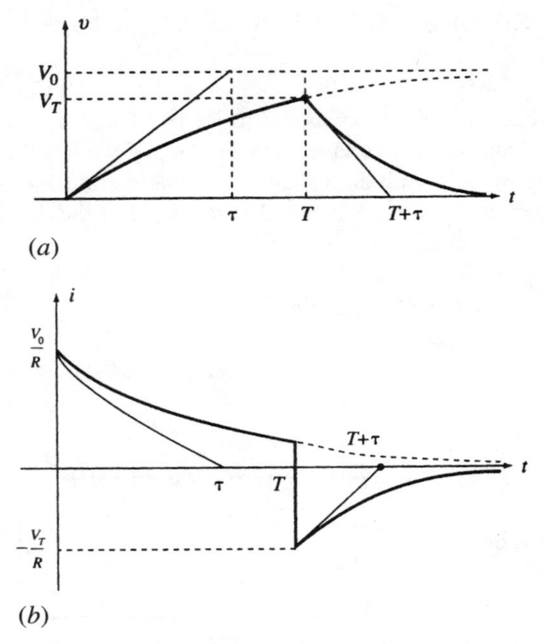

Figure 4-17 Response of *RC* circuit to a pulse

Example 4.10 In the circuit of Figure 4-16, let $R = 1\ \text{k}\Omega$ and $C = 1\ \mu\text{F}$ and let the voltage source be a 1-V pulse of duration 1 ms. Find v and i.

Solution: We use (7) through (11) with the time constant $\tau = RC = 1$ ms. The pulse response for this circuit is

$$v(t) = 1 - e^{-1000t}\ \text{V}, \quad 0 < t < 10^{-3}\ \text{s}$$

$$i(t) = 1e^{-1000t}\ \text{mA}, \quad 0 < t < 10^{-3}\ \text{s}$$

$$V_T = 1 - e^{-1} = 0.632\ \text{V}$$

$$v(t) = 0.632e^{-1000(t-10^{-3})}\ \text{V}, t > 10^{-3}\ \text{s}$$

$$i(t) = -0.632e^{-1000(t-10^{-3})}\ \text{mA}, t > 10^{-3}\ \text{s}$$

Response of a First-Order Circuit to an Impulse

A narrow pulse can be modeled as an impulse with the area under the pulse indicating its strength. Impulse response is a useful tool in analysis and synthesis of circuits. It may be derived in several ways including: take the limit of the response to a narrow pulse (*limit approach*); take the derivative of the step response; solve the differential equation directly.

The impulse response is often designated by $h(t)$.

Example 4.11 Find the limits of v and i of the circuit from Figure 4-16 for a voltage pulse of unit area as the pulse duration is decreased to zero.

Solution: We use the pulse responses in (7) through (11) with $V_0 = 1/T$ and find their limits as the T approaches zero. From (9) we have

$$\lim_{T\to 0} V_T = \lim_{T\to 0}(1 - e^{-T/RC})/T = 1/RC$$

From (10) and (11) we have

For $t < 0$, $\quad h_v = 0$ and $h_i = 0$

For $0^- < t < 0^+$, $\quad 0 \le h_v \le \dfrac{1}{RC}$ and $h_i = \dfrac{1}{R}\delta(t)$

For $t > 0$, $\quad h_v(t) = \dfrac{1}{RC}e^{-t/RC}$ and $h_i(t) = -\dfrac{1}{R^2C}e^{-t/RC}$

Therefore,

$$h_v(t) = \frac{1}{RC}e^{-t/RC}u(t) \text{ and } h_i(t) = \frac{1}{R}\delta(t) - \frac{1}{R^2C}e^{-t/RC}u(t)$$

Example 4.12 Find the impulse response of the *RC* circuit in Figure 4-16 by taking the derivatives of its unit step response.

Solution: A unit impulse may be considered the derivative of a unit step. Based on the properties of linear differential equations with constant coefficients, we can take the time derivative of the step response to find the impulse response. The unit step responses of an *RC* circuit were found in (5) and (6) to be

$$v(t) = (1 - e^{-t/RC})\, u(t) \quad \text{and} \quad i(t) = \frac{1}{R} e^{-t/RC}\, u(t)$$

We find the unit impulse responses by taking the derivatives of the step responses. Thus

$$h_v(t) = \frac{1}{RC} e^{-t/RC} u(t) \text{ and } h_i(t) = \frac{1}{R}\delta(t) - \frac{1}{R^2C} e^{-t/RC} u(t)$$

Important Things to Remember

- ✔ Periodic functions repeat every T seconds, where T is the period and $f = 1/T$ is the frequency in hertz.
- ✔ The unit impulse function $\delta(t)$ is the derivative of the unit step function $u(t)$.
- ✔ The transient solution of first-order *RC* and *RL* circuits will generally involve some type of exponential solution.
- ✔ For a linear circuit, the impulse response $h(t)$ is the derivative of the unit step response.

Chapter 5

SINUSOIDAL STEADY-STATE CIRCUIT ANALYSIS

IN THIS CHAPTER:

- ✔ *Element Responses*
- ✔ *Phasors*
- ✔ *Impedance and Admittance*
- ✔ *Phasor Analysis Methods*

Element Responses

This chapter will concentrate on the steady-state response of circuits driven by sinusoidal sources. The response will also be sinusoidal.

The voltage–current relationships for the single elements R, L, and C were examined in Chapter 1 and summarized in Table 1-4. In this chapter, the functions v and i will be sines or cosines with the argument ωt. ω is the angular frequency and has the unit rad/s. Also, $\omega = 2\pi f$, where f is the frequency in Hz.

Consider an inductance L with $i = I\cos(\omega t + 45°)$ A [see Figure 5-1(a)].

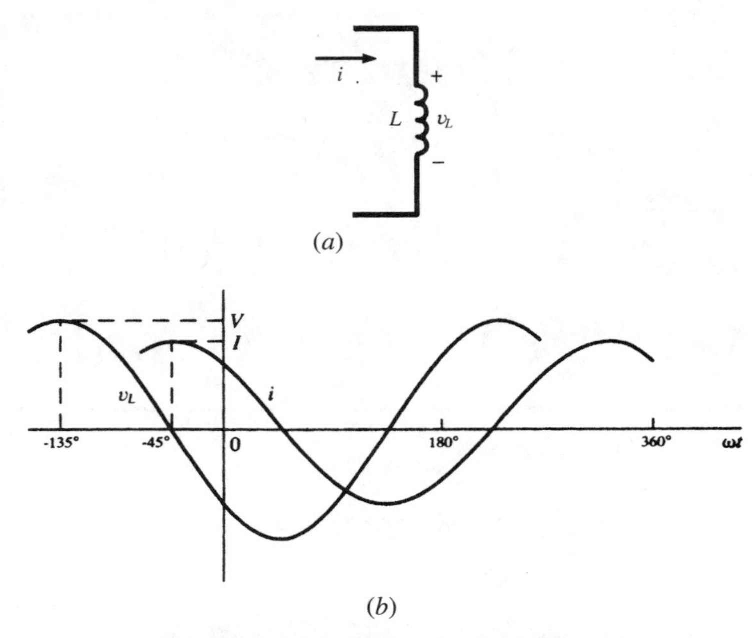

Figure 5-1 Sinusoidal current and voltage for an inductor

The voltage is

$$v_L = L\frac{di}{dt} = \omega LI[-\sin(\omega t + 45°)] = \omega LI\cos(\omega t + 135°)]\text{ V}$$

A comparison of v_L and i shows that the current *lags* the voltage by 90° or $\pi/2$ radians. The functions are sketched in Figure 5-1(b).

The current function *i* is to the right of v_L, and since the horizontal scale is ωt, events displaced to the right occur later in time. This illustrates that ***i* lags *v*** for an inductor.

The horizontal scale is in radians, but note that it is also marked in degrees (–135°, 180°, etc.). This is a case of mixed units just as with $\omega t + 45°$. It is not mathematically correct but is the accepted practice in circuit analysis. The vertical scale indicates two different quantities, that is v and i, so there should be two scales rather than one.

In Table 5-1, the responses of the three basic circuit elements are shown for applied current $i = I \cos \omega t$ and voltage $v = V \cos \omega t$.

Table 5-1

	$i = I \cos \omega t$	$v = V \cos \omega t$
i, v_R, R (+ / –)	$v_r = RI \cos \omega t$	$i_R = \frac{V}{R} \cos \omega t$
i, v_L, L (+ / –)	$v_L = \omega LI \cos(\omega t + 90°)$	$i_L = \frac{V}{\omega L} \cos(\omega t - 90°)$
i, v_C, C (+ / –)	$v_C = \frac{I}{\omega C} \cos(\omega t - 90°)$	$i_C = \omega CV \cos(\omega t + 90°)$

The relations also show that for a resistance R, the voltage v and current i are in phase. For a capacitance C, ***i* leads *v*** by 90°.

Example 5.1 The RL series circuit shown in Figure 5-2 has a current $i = I \sin \omega t$. Obtain the voltage v across the two circuit elements and sketch v and i.

Solution:

$$v_R = RI \sin \omega t$$

$$v_L = L\frac{di}{dt} = \omega LI \sin(\omega t + 90°)$$

$$v = v_R + v_L = RI \sin \omega t + \omega LI \sin(\omega t + 90°)$$

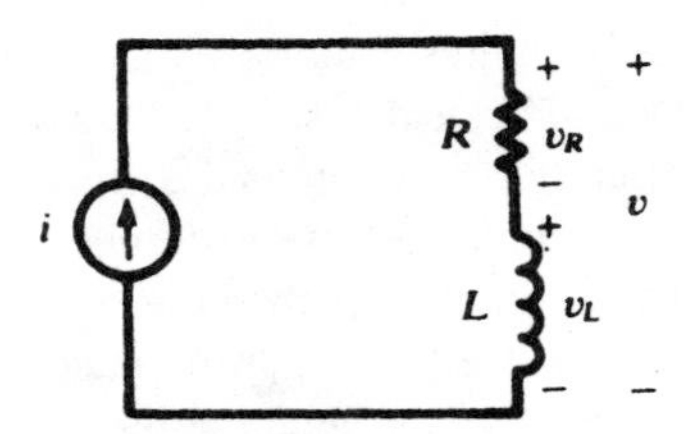

Figure 5-2 Series *RL* example circuit

Since the current is a sine function and, from a trig identity,

$$v = V\sin(\omega t + \theta) = V\sin\omega t\cos\theta + V\cos\omega t\sin\theta \tag{1}$$

we have from the above

$$v = RI\sin\omega t + \omega LI\sin\omega t\cos 90^\circ + \omega LI\cos\omega t\sin 90^\circ \tag{2}$$

Equating coefficients of like terms in (1) and (2),

$$V\sin\theta = \omega LI \text{ and } V\cos\theta = RI$$

Then

$$v = I\sqrt{R^2 + (\omega L)^2}\,\sin[\omega t + \tan^{-1}(\omega L / R)]$$

and

$$V = I\sqrt{R^2 + (\omega L)^2} \text{ and } \theta = \tan^{-1}(\omega L / R)$$

The functions i and v are sketched in Figure 5-3. The phase angle θ, the angle by which i lags v, lies within the range $0^\circ \le \theta \le 90^\circ$, with the limiting values attained for $\omega L << R$ and $\omega L >> R$, respectively.

If the circuit had an applied voltage $v = V\sin\omega t$, the resulting current would be

$$i = \frac{V}{\sqrt{R^2 + (\omega L)^2}}\sin(\omega t - \theta)$$

where, as before, $\theta = \tan^{-1}(\omega L/R)$.

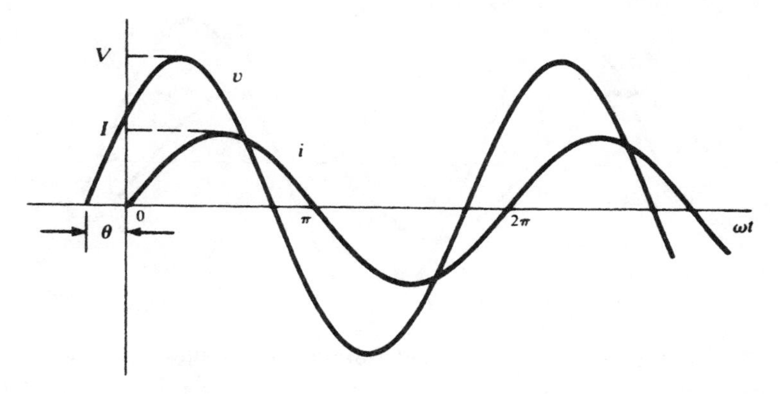

Figure 5-3 Current and voltage for Example 5.1

Example 5.2 If the current driving a series *RC* circuit is given by $i = I\sin\omega t$, obtain the total voltage and current across the two elements.

Solution

$$v_R = RI\sin\omega t$$

$$v_C = \frac{1}{\omega C}\sin(\omega t - 90°)$$

$$v = v_R + v_C = V\sin(\omega t - \theta)$$

where

$$V = I\sqrt{R^2 + (1/\omega C)^2} \text{ and } \theta = \tan^{-1}(1/\omega CR)$$

The negative phase angle shifts v to the right of the current i. Consequently, i leads v for a series *RC* circuit. The phase angle is constrained to the range of $0° \le \theta \le 90°$. For $(1/\omega C) << R$, the angle $\theta \approx 0°$ and for $(1/\omega C) >> R$, the angle $\theta \approx 90°$. See Figure 5-4.

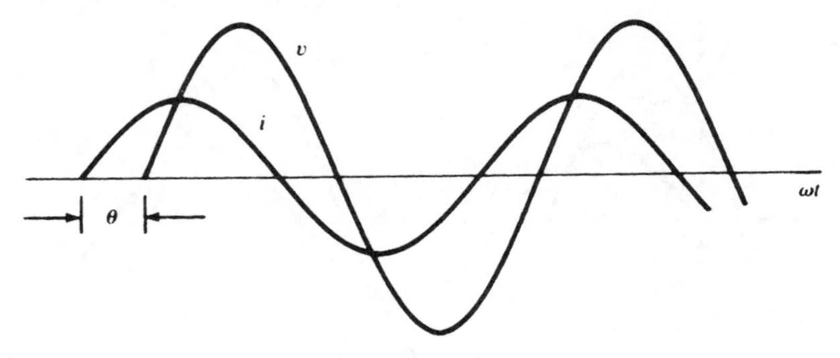

Figure 5-4 Current and voltage for Example 5.1

Phasors

A brief look at the voltage and current sinusoids in the preceding examples shows that the amplitudes and phase differences are the two principal concerns. A directed line segment, or *phasor*, such as that shown rotating in a counterclockwise direction at a constant angular velocity ω (rad/s) in Figure 5-5, has a projection on the horizontal that is a cosine function. The length of the phasor or its *magnitude* is the amplitude or maximum value of the cosine function. The angle between two positions of the phasor is the *phase difference* between the corresponding points on the cosine function.

Remember

Throughout this book, *phasors will be defined from the cosine function.* If a voltage or current is expressed as a sine, it will be changed to a cosine by subtracting 90° from the phase.

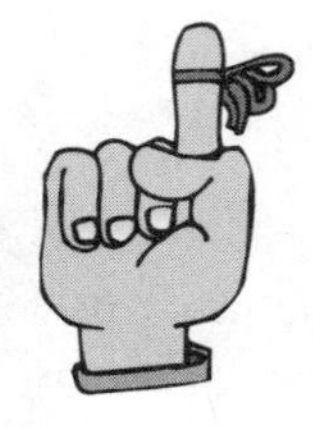

Consider the examples shown in Table 5-2. Observe that the phasors, which are directed line segments and vectorial in nature, are indicated by

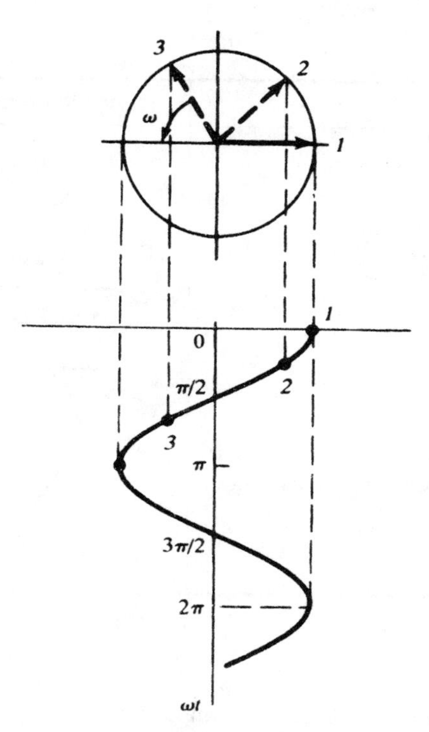

Figure 5-5 Relation of a phasor to a sinusoid

boldface capitals, for example, **V** and **I**. The phase angle of the cosine function is the angle on the phasor. The phasor diagrams here and all that follow may be considered as a snapshot of the counterclockwise-rotating directed line segment taken at $t = 0$. The frequency f (Hz) and ω (rad/s) generally do not appear but they should be kept in mind, since they are implicit in any sinusoidal steady-state problem.

Example 5.3 A series combination of $R = 10\ \Omega$ and $L = 20$ mH has a current $i = 5.0\cos(500t + 10°)$ A. Obtain the voltages v and **V**, the phasor current **I**, and sketch the phasor diagram.

Table 5-2

Function	Phasor Representation
$v = 150\cos(500t + 45^\circ)$ (V)	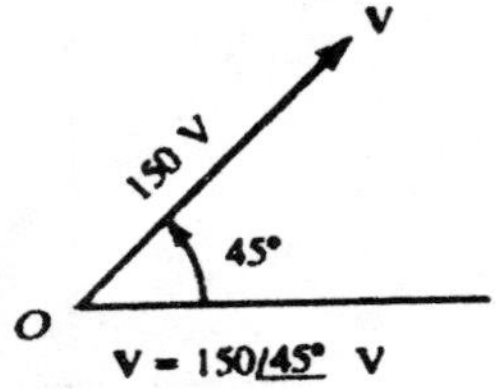 $\mathbf{V} = 150\angle 45^\circ$ V
$i = 3.0\sin(2000t + 30^\circ)$ (mA) $= 3.0\cos(2000t - 60^\circ)$ (mA)	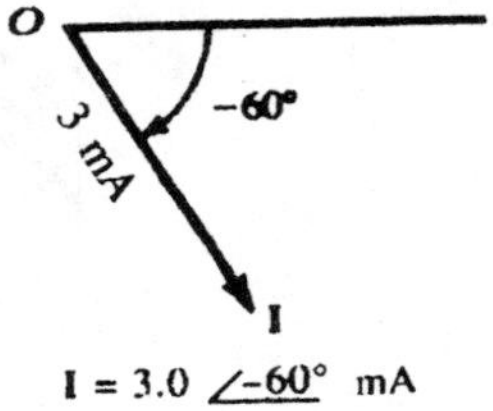 $\mathbf{I} = 3.0\angle -60^\circ$ mA

Solution: Using the methods of Example 5.1,

$$v_R = 50.0\cos(500t + 10^\circ)\ \text{V}$$

$$v_L = L\frac{di}{dt} = 50.0\cos(500t + 100^\circ)\ \text{V}$$

$$v = v_R + v_L = 70.7\cos(500t + 55^\circ)\ \text{V}$$

The corresponding phasors are

$$\mathbf{I} = 5.0\angle 10^\circ\ \text{A and } \mathbf{V} = 70.7\angle 55^\circ\ \text{V}$$

The phase angle of 45° can be seen in the time-domain graphs of Figure 5-6(*a*), and the phasor diagram with **I** and **V** shown in Figure 5-6(*b*).

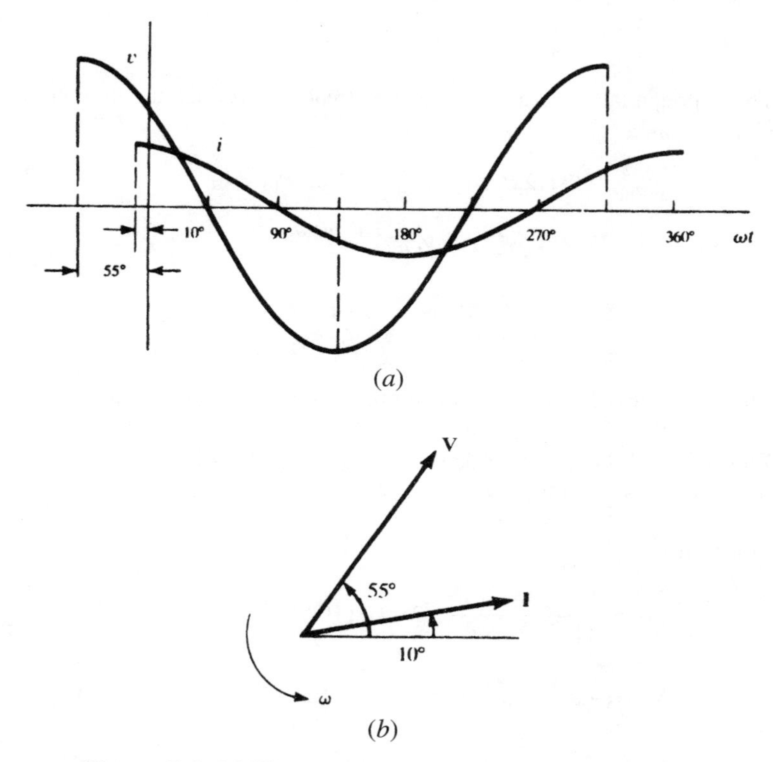

Figure 5-6 (*a*) Time domain and (*b*) phasor diagrams

Phasors can be treated as complex numbers. When the horizontal axis is identified as the real axis of a complex plane, the phasors become complex numbers and the usual rules apply (see Appendix for discussion of complex numbers).

In view of *Euler's identity*, there are three equivalent notations for a phasor:

Polar form	$\mathbf{V} = V\angle\theta$
Rectangular form	$\mathbf{V} = V(\cos\theta + j\sin\theta)$
Exponential form	$\mathbf{V} = Ve^{j\theta}$

The cosine expression may also be written as

$$v = V\cos(\omega t + \theta) = \text{Re}[\mathbf{V}e^{j\omega t}]$$

The exponential form suggests how to treat the product and quotient of phasors. Since $(V_1 e^{j\theta_1})(V_2 e^{j\theta_2}) = V_1 V_2 e^{j\theta_1 + \theta_2}$,

$$(V_1\angle\theta_1)(V_2\angle\theta_2) = V_1 V_2 \angle(\theta_1 + \theta_2)$$

and, since $(V_1 e^{j\theta_1}) / (V_2 e^{j\theta_2}) = (V_1 / V_2)\, e^{j\theta_1 - \theta_2}$,

$$\frac{V_1\angle\theta_1}{V_2\angle\theta_2} = \frac{V_1}{V_2}\angle(\theta_1 - \theta_2)$$

The rectangular form is used in summing or subtracting phasors.

Example 5.4 Given $\mathbf{V}_1 = 25.0\angle 143.13°$ and $\mathbf{V}_2 = 11.2\angle 26.57°$, find the ratio $\mathbf{V}_1/\mathbf{V}_2$ and the sum $\mathbf{V}_1 + \mathbf{V}_2$.

Solution:

$$\frac{\mathbf{V}_1}{\mathbf{V}_2} = \frac{25.0\angle 143.13°}{11.2\angle 26.57°} = 2.23\angle 116.56° = -1.00 + j1.99$$

$$\mathbf{V}_1 + \mathbf{V}_2 = (-20 + j15) + (10 + j5) = -10 + j20 = 23.36\angle 116.57°$$

Impedance and Admittance

A sinusoidal voltage or current applied to a passive *RLC* circuit produces a sinusoidal response. With time functions, such as $v(t)$ and $i(t)$, the circuit is said to be in the *time domain*, Figure 5-7(a). When the circuit is analyzed using phasors, it is said to be in the *frequency domain*, Figure 5-7(b). The voltage and current may be written, respectively,

$$v(t) = V\cos(\omega t + \theta) = \text{Re}[\mathbf{V}e^{j\omega t}] \text{ and } \mathbf{V} = V\angle\theta$$

$$i(t) = I\cos(\omega t + \theta) = \text{Re}[\mathbf{I}e^{j\omega t}] \text{ and } \mathbf{I} = I\angle\theta$$

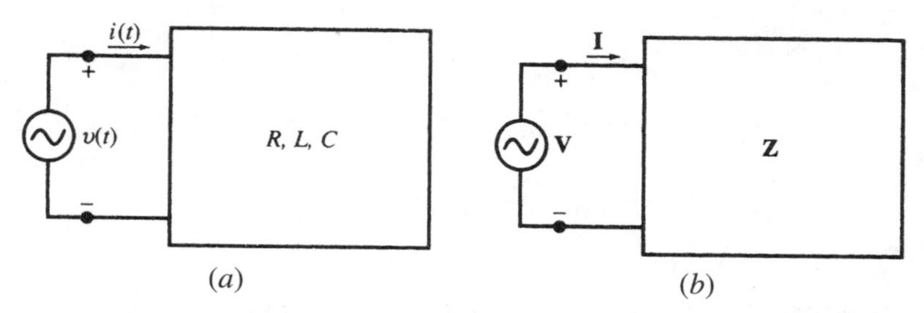

Figure 5-7 (*a*) Time domain and (*b*) frequency domain *RLC* circuits

The ratio of phasor voltage **V** to phasor current **I** is defined as *impedance* **Z**, that is, **Z** = **V/I** Ω.

The reciprocal of impedance is called *admittance* **Y**, so that **Y** = 1/**Z** (S) where 1 Siemen (S) = 1 Ω^{-1}. **Y** and **Z** are complex numbers.

When impedance is written in Cartesian form, the real part is the resistance R and the imaginary part is the *reactance* X. The sign on the imaginary part may be positive or negative.

When the sign is positive, X is called the *inductive reactance*. When the sign is negative, X is called the *capacitive reactance*.

When the admittance is written in Cartesian form, the real part is *admittance* G and the imaginary part is *susceptance* B.

A positive sign on B indicates a *capacitive susceptance*. A negative sign on B indicates an *inductive susceptance*.

Thus,

$$\mathbf{Z} = R + jX_L \quad \text{and} \quad \mathbf{Z} = R - jX_C$$

$$\mathbf{Y} = G - jB_L \quad \text{and} \quad \mathbf{Y} = G + jB_C$$

Example 5.5 The phasor voltage across the terminals of a network such as that shown in Figure 5-7(*b*) is 100 ∠45° V and the resulting current is 5.0 ∠15° A. Find the equivalent impedance and admittance.

Solution:

$$\mathbf{Z} = \frac{\mathbf{V}}{\mathbf{I}} = \frac{100\angle 45^\circ}{5\angle 15^\circ} = 20.0\angle 30^\circ = 17.32 + j10\ \Omega$$

$$\mathbf{Y} = \frac{\mathbf{I}}{\mathbf{V}} = \frac{1}{\mathbf{Z}} = 0.05\angle -30^\circ = (4.33 - j2.5)\times 10^{-2}\ \text{S}$$

Thus, $R = 17.32\ \Omega$, $X_L = 10\ \Omega$, $G = 4.33\times 10^{-2}$ S, and $B_L = 2.5\times 10^{-2}$ S.

The relation $\mathbf{V} = \mathbf{IZ}$ (in the frequency domain) is identical to Ohm's law, $v = iR$, for a resistive network (in the time domain). Therefore, impedances combine exactly like resistances:

Impedances in series: $\mathbf{Z}_{eq} = \mathbf{Z}_1 + \mathbf{Z}_2 + \ldots$

Impedances in parallel: $\dfrac{1}{\mathbf{Z}_{eq}} = \dfrac{1}{\mathbf{Z}_1} + \dfrac{1}{\mathbf{Z}_2} + \ldots$

Two impedances in parallel: $\mathbf{Z}_{eq} = \dfrac{\mathbf{Z}_1\mathbf{Z}_2}{\mathbf{Z}_1 + \mathbf{Z}_2}$

In an *impedance diagram*, a point in the right half of the complex plane represents an impedance $\mathbf{Z}$. Figure 5-8 shows two impedances $\mathbf{Z}_1$, in the first quadrant, exhibits inductive reactance, while $\mathbf{Z}_2$, in the fourth quadrant, exhibits capacitive reactance. Their series equivalent, $\mathbf{Z}_1 + \mathbf{Z}_2$, is obtained by vector addition, as shown. Note that the "vectors" are shown

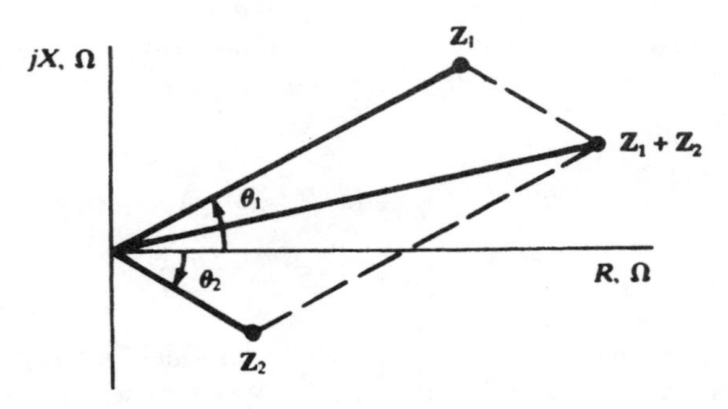

Figure 5-8 Impedance diagram

without arrowheads, in order to distinguish these complex numbers from phasors.

Replacing **Z** by 1/**Y** in the formulas gives

Admittance in series: $\frac{1}{\mathbf{Y}_{eq}} = \frac{1}{\mathbf{Y}_1} + \frac{1}{\mathbf{Y}_2} + \ldots$

Admittance in parallel: $\mathbf{Y}_{eq} = \mathbf{Y}_1 + \mathbf{Y}_2 + \ldots$

Thus, series circuits are easiest treated in terms of impedance and parallel circuits are easiest treated in terms of admittance.

Figure 5-9, an *admittance diagram*, is analogous to Figure 5-8 for impedance. Shown are an admittance $\mathbf{Y}_1$ having capacitive susceptance and an admittance $\mathbf{Y}_2$ having inductive susceptance, together with their vector sum $\mathbf{Y}_1 + \mathbf{Y}_2$, which is the admittance of a parallel combination of $\mathbf{Y}_1$ and $\mathbf{Y}_2$.

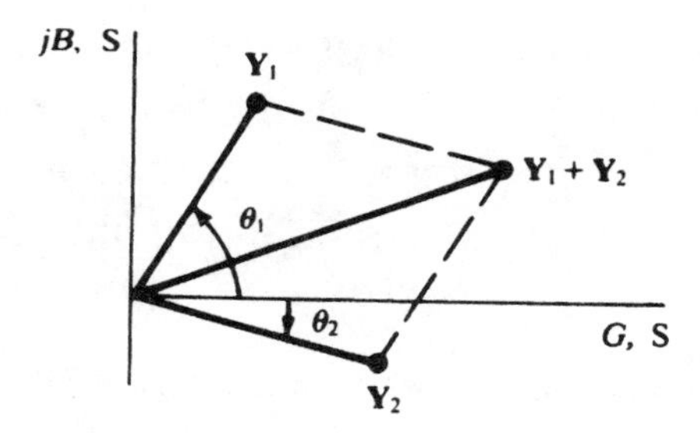

Figure 5-9 Admittance diagram

Phasor Analysis Methods

The analogy between impedance in the frequency domain and resistance in the time domain, discussed in Chapter 2, implies the following results:

1. Impedances in series divide the total voltage in the ratio of impedances:

$$\mathbf{V}_r = \frac{\mathbf{Z}_r}{\mathbf{Z}_T}\mathbf{V}_T$$

2. Impedances in parallel divide the total current in the inverse ratio of impedances (direct ratio of admittances)

$$\mathbf{I}_r = \frac{\mathbf{Z}_{eq}}{\mathbf{Z}_r}\mathbf{I}_T = \frac{\mathbf{Y}_r}{\mathbf{Y}_T}\mathbf{I}_T$$

In fact, many of the time-domain analysis methods directly translate to the frequency domain using phasors and impedance and admittance. This includes the mesh current method, the node voltage method, Thévenin's and Norton's theorems, superposition, and more.

You Need to Know 

In order to use phasors with the above-mentioned methods with circuits having multiple sources, all sources must have the same frequency *f*. If the sources are of different frequencies, then the analysis must be performed in the time domain.

Example 5.6 Replace the active network in Figure 5-10 at terminals *ab* with a Thévenin equivalent.

Solution:

$$\mathbf{Z}' = j5 + \frac{5(3+j4)}{5+3+j4} = 2.5 + j6.25\ \Omega$$

The open circuit voltage $\mathbf{V}'$ at terminals *ab* is the voltage across the $3 + j4\Omega$ impedance:

$$\mathbf{V}' = 10\angle 0^\circ \frac{3+j4}{8+j4} = 5.59\angle 26.56^\circ\ \mathrm{V}$$

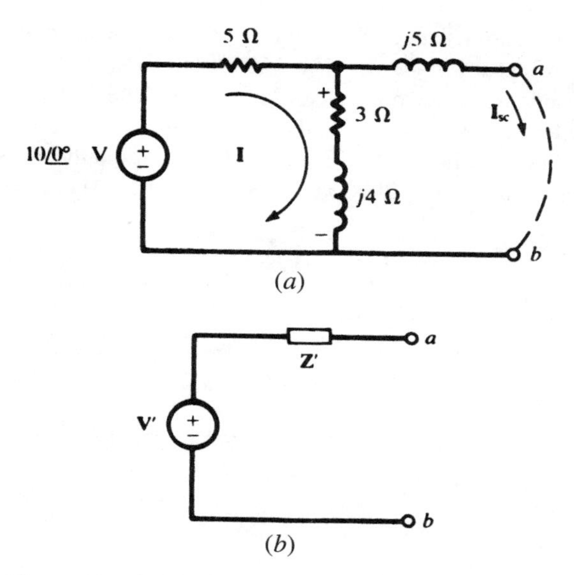

Figure 5-10 Thévenin equivalent for Example 5.6

Example 5.7 For the circuit of Example 5.6, obtain a Norton equivalent circuit (Figure 5-11).

Solution: At terminals *ab*, $\mathbf{I}_{sc}$ is the Norton current $\mathbf{I}'$. By current division,

$$\mathbf{I}' = \frac{10\angle 0^\circ}{5 + \dfrac{5(3+j4)}{5+3+j4}}\left(\frac{3+j4}{3+j9}\right) = 0.830\angle -41.63^\circ \text{ A}$$

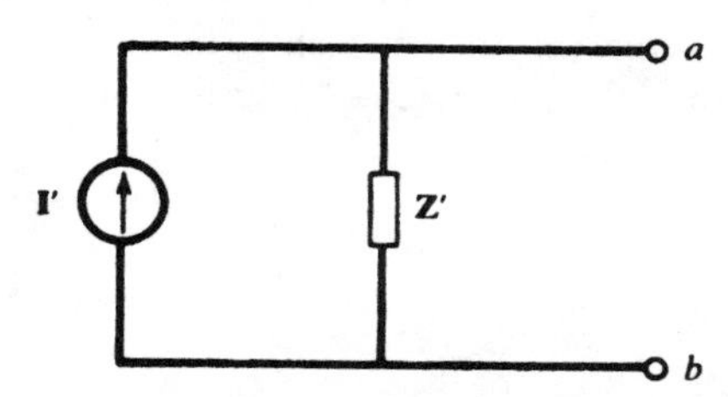

Figure 5-11 Norton equivalent circuit for Example 5.7

Example 5.8 A practical coil is connected in series between two voltage sources $v_1 = 5\cos\omega_1 t$ V and $v_2 = 10\cos(\omega_2 t + 60°)$ V such that the sources share the same reference node (see Figure 5-12). The voltage difference across the terminals of the coil is therefore $v_1 - v_2$. The coil is modeled by a 5-mH inductor in series with a 10-Ω resistor. Find the current $i(t)$ in the coil for (*a*) $\omega_1 = \omega_2 = 2000$ rad/s and (*b*) $\omega_1 = 2000$ rad/s, $\omega_2 = 2\omega_1$.

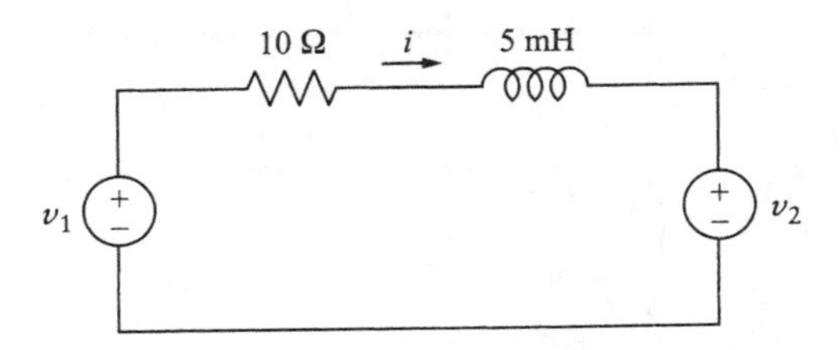

Figure 5-12 Practical coil example

Solution:

(*a*) The impedance of the coil is $R + j\omega L = 10 + j10 = 10\sqrt{2}\angle 45°$ Ω. The phasor voltage between its terminals is given by the difference $\mathbf{V} = \mathbf{V}_2 - \mathbf{V}_1 = 5 - 10\angle 60° = -j5\sqrt{3}$ V. The current is

$$\mathbf{I} = \frac{\mathbf{V}}{\mathbf{Z}} = \frac{-j5\sqrt{3}}{10\sqrt{2}\angle 45°} = \frac{-j8.66}{14.14\angle 45°} = 0.61\angle -135° \text{ A}$$

$$i(t) = 0.61\cos(2000t - 135°) \text{ A}$$

(*b*) Because the coil has different impedances at $\omega_1 = 2000$ rad/s and $\omega_2 = 4000$ rad/s, the current may be represented in the time domain only. By applying superposition, we get $i = i_1 - i_2$, where i_1 and i_2 are currents due to v_1 and v_2, respectively.

$$\mathbf{I}_1 = \frac{\mathbf{V}_1}{\mathbf{Z}_1} = \frac{5}{10 + j10} = 0.35\angle -45° \text{ A} \Rightarrow i_1(t) = 0.35\cos(2000t - 45°) \text{ A}$$

$$\mathbf{I}_2 = \frac{\mathbf{V}_2}{\mathbf{Z}_2} = \frac{10\angle 60°}{10 + j20} = 0.45\angle -3.4° \text{ A} \Rightarrow i_2(t) = 0.45\cos(4000t - 3.4°) \text{ A}$$

$$i(t) = i_1(t) - i_2(t) = 0.35\cos(2000t - 45°) - 0.45\cos(4000t - 3.4°) \text{ A}$$

Important Things to Remember

- ✔ In a resistor, the current and the voltage are in phase.
- ✔ In an inductor, the current lags the voltage by 90°.
- ✔ In a capacitor, the current leads the voltage by 90°.
- ✔ Phasor analysis is referred to as *frequency-domain analysis.*
- ✔ By defining impedance and admittance, the circuit analysis techniques from the time domain generally apply in the frequency domain.
- ✔ However, if there are multiple sources at different frequencies, the circuit must be analyzed by superposing the individual time-domain solutions.

Chapter 6
AC Power

IN THIS CHAPTER:

- ✔ *Power in the Time Domain*
- ✔ *Sinusoidal Steady-State Power*
- ✔ *Complex Power*

Power in the Time Domain

The *instantaneous power* entering a two-terminal circuit N (Figure 6-1) is defined by

$$p(t) = v(t)i(t) \tag{1}$$

where $v(t)$ and $i(t)$ are terminal voltage and current, respectively.

If $p(t)$ is positive, energy is delivered to the circuit. If $p(t)$ is negative, energy is returned from the circuit to the source.

In this chapter, we consider periodic currents and voltages, with emphasis on the sinusoidal steady-state *RLC* circuits. Since the storage ca-

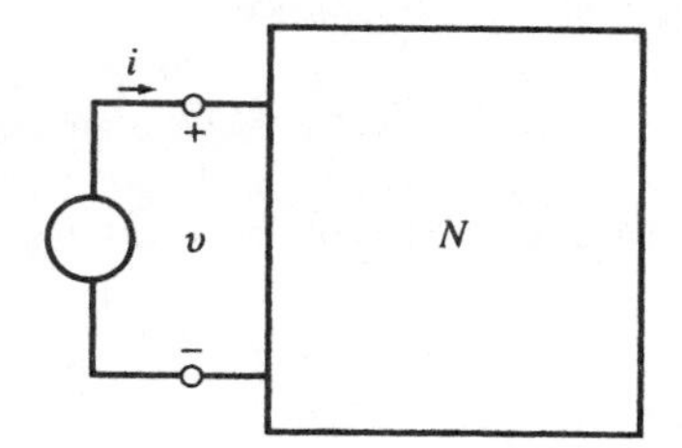

Figure 6-1 Two-terminal circuit

pacity of an inductor or a capacitor is finite, these passive elements cannot continue receiving energy without returning it. Therefore, in steady state and during each cycle, all of the energy received by an inductor or capacitor is returned. The energy received by a resistor is, however, dissipated in the form of thermal, mechanical, chemical, and/or electromagnetic energies. The net energy flow to a passive circuit during one cycle is, therefore, positive or zero.

Example 6.1 Figure 6-2(*a*) shows the graph of a current in a resistor of 1 kΩ. Find and plot the instantaneous power $p(t)$.

Solution: From $v = Ri$, we have

$$p(t) = vi = Ri^2 = 1000 \times 10^{-6} = 10^{-3}\ \text{W} = 1\ \text{mW}$$

See Figure 6-2(*b*).

Sinusoidal Steady-State Power

A sinusoidal voltage $v = V_m \cos \omega t$, applied across an impedance $\mathbf{Z} = |Z| \angle \theta$, establishes a current $i = I_m \cos(\omega t - \theta)$. The power delivered to the impedance at time t is

$$\begin{aligned} p(t) &= vi = V_m I_m \cos \omega t \cos(\omega t - \theta) = \frac{1}{2} V_m I_m [\cos \theta + \cos(2\omega t - \theta)] \\ &= V_{eff} I_{eff} [\cos \theta + \cos(2\omega t - \theta)] \qquad (2) \\ &= V_{eff} I_{eff} \cos \theta + V_{eff} I_{eff} \cos(2\omega t - \theta) \end{aligned}$$

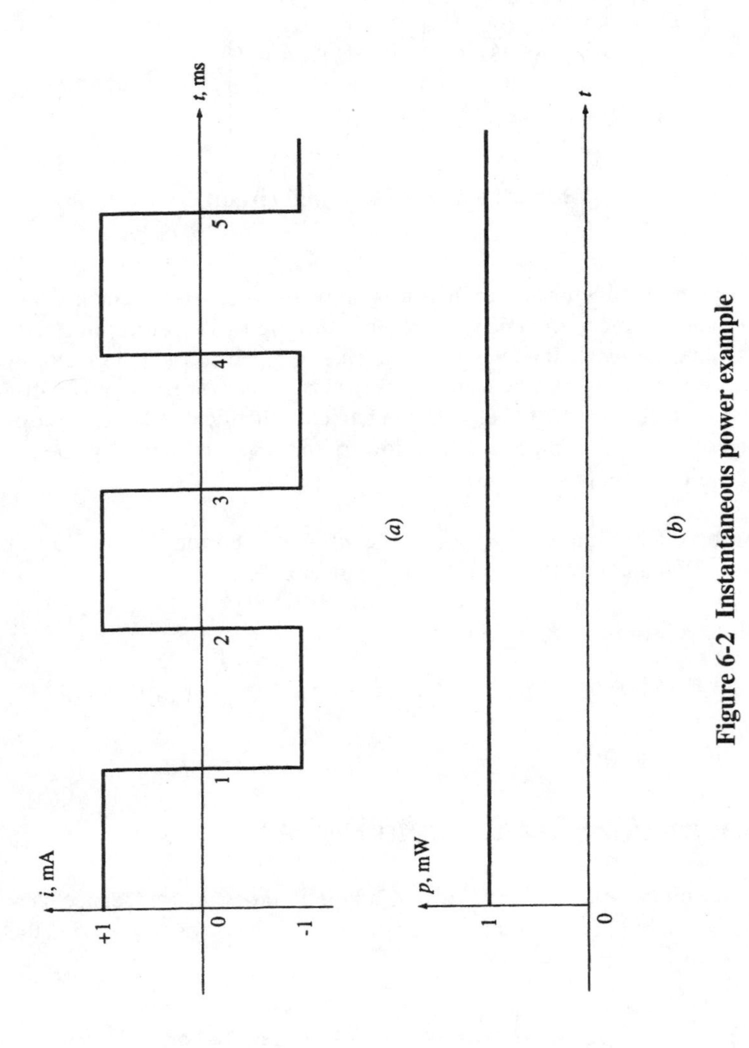

Figure 6-2 Instantaneous power example

where $V_{eff} = V_m / \sqrt{2}$, $I_{eff} = I_m / \sqrt{2}$, and $I_{eff} = V_{eff} / |\mathbf{Z}|$. The instantaneous power in (2) consists of a sinusoidal component $V_{eff} I_{eff} \cos(2\omega t - \theta)$ plus a constant value $V_{eff} I_{eff} \cos\theta$ that becomes the average power P_{avg}. This is illustrated in Figure 6-3. During a portion of one cycle, the instantaneous power is positive, which indicates that the power flows into the load. During the rest of the cycle, the instantaneous power may be negative, which indicates that the power flows out of the load. The net flow of power during one cycle is, however, nonnegative and is called the *average power*.

Example 6.2 A voltage $v = 140 \cos \omega t$ V is connected across an impedance $\mathbf{Z} = 5\angle{-60°}$. Find $p(t)$.

Solution: The voltage v results in a current $i = 28\cos(\omega t + 60°)$ A. Then,

$$p(t) = vi = 140(28) \cos \omega t \cos(\omega t + 60°) = 980 + 1960 \cos(2\omega t + 60°) \text{ W}$$

The instantaneous power has a constant component of 980 W and a sinusoidal component with twice the frequency of the source. The plot of p vs. t is similar to that in Figure 6-3 with $\theta = -\pi/3$ rad.

You Need to Know!

The net or average power $P_{avg} = \langle p(t) \rangle$ entering a load during one period is called the *real power*.

Since the average of $\cos(2\omega t - \theta)$ over one period is zero, from (2) we get

$$P_{avg} = V_{eff} I_{eff} \cos\theta \tag{3}$$

If $\mathbf{Z} = R + jX = |\mathbf{Z}|\angle\theta$, then $\cos\theta = R/|\mathbf{Z}|$ and P_{avg} may be expressed by

$$P_{avg} = V_{eff} I_{eff} \frac{R}{|\mathbf{Z}|} \tag{4}$$

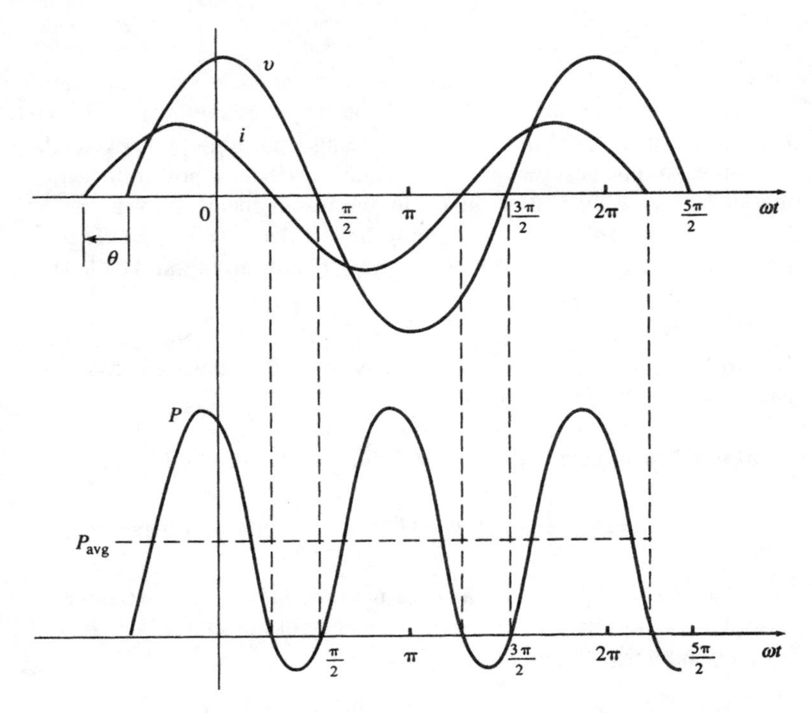

Figure 6-3 Voltage, current, and power relationships

or

$$P_{avg} = \frac{V_{eff}^2}{|\mathbf{Z}|^2} R \tag{5}$$

or

$$P_{avg} = RI_{eff}^2 \tag{6}$$

The average power is nonnegative. It depends on V, I, and the phase angle between them. When V_{eff} and I_{eff} are given, P is maximum for $\theta = 0$. The ratio of P_{avg} to $V_{eff} I_{eff}$ is called the *power factor* pf.

From (3), the ratio is equal to $\cos\theta$ and so

$$\text{pf} = \frac{P_{avg}}{V_{eff} I_{eff}}, 0 \le \text{pf} \le 1 \tag{7}$$

The subscript "*avg*" in the average power is often omitted. In the remainder of this chapter, P will denote the average power.

Example 6.3 Find P delivered from a sinusoidal voltage source with $V_{eff} = 110$ V to an impedance of $\mathbf{Z} = 10 + j8\Omega$. Find the power factor.

Solution:

$$\mathbf{Z} = 10 + j8\Omega = 12.81\angle 38.7°\Omega$$

$$I_{eff} = \frac{V_{eff}}{\mathbf{Z}} = \frac{110}{12.81\angle 38.7°} = 8.59\angle -38.7° \text{ A}$$

$$P = V_{eff} I_{eff} \cos\theta = 110(8.59)(\cos 38.7°) = 737.43 \text{ W}$$

$$\text{pf} = \cos 38.7° = 0.78$$

Another solution for the average power uses $|\mathbf{Z}|^2 = 100 + 64 = 164$. Then,

$$P = V_{eff}^2 R / |\mathbf{Z}|^2 = 110^2(10)/164 = 737.43 \text{ W}$$

If a passive network contains inductors, capacitors, or both, a portion of energy entering it during one cycle is stored and then returned to the source. During the period of energy return, the power is negative. The power involved in this exchange is called *reactive* or *quadrature power*. Although the net effect of reactive power is zero, it degrades the operation of power systems. Reactive power, indicated by Q, is defined as

$$Q = V_{eff} I_{eff} \sin\theta \tag{8}$$

If $\mathbf{Z} = R + jX = |\mathbf{Z}|\angle\theta$, then $\sin\theta = X/|\mathbf{Z}|$ and Q may be expressed by

$$Q = V_{eff} I_{eff} \frac{X}{|\mathbf{Z}|} \tag{9}$$

$$Q = \frac{V_{eff}^2}{|\mathbf{Z}|^2} X \tag{10}$$

$$Q = X I_{eff}^2 \tag{11}$$

The unit of reactive power is the *volt-amperes reactive* (var).

The reactive power Q depends on V, I, and the phase angle between them. It is the product of the voltage and that component of the current that is 90° out of phase with the voltage.

Q is zero for $\theta = 0°$. This occurs for a purely resistive load, when **V** and **I** are in phase. When the load is purely reactive, $|\theta| = 90°$ and Q attains its maximum value.

While P is always nonnegative, Q can assume positive values (for an inductive load where current lags the voltage) or negative values (for a capacitive load where current leads the voltage). It is also customary to specify Q by its magnitude and load type. For example, 100-kvar inductive means $Q = 100$ kvar and 100-kvar capacitive indicates $Q = -100$ kvar.

Example 6.4 The voltage and current across a load are given by $V_{eff} = 110$ V and $I_{eff} = 20\angle -50°$A. Find P and Q.

Solution:

$$P = 110(20\cos 50°) = 1414 \text{ W}$$

$$Q = 110(20\sin 50°) = 1685 \text{ var}$$

AC power in resistors, inductors, and capacitors is summarized in Table 6-1. The last column of Table 6-1 is $S = VI$ where S is called *apparent power*. S is discussed in the next section in more detail.

Example 6.5 Find the power delivered from a sinusoidal source to a resistor R.

Table 6-1

$v = (V\sqrt{2})\cos\omega t \qquad \mathbf{V}_{eff} = V\angle 0^\circ$
$i = (I\sqrt{2})\cos(\omega t - \theta) \qquad \mathbf{I}_{eff} = I\angle -\theta^\circ$
$P = VI\cos\theta$, $Q = VI\sin\theta$ and $S = VI$ (apparent power)

	$\mathbf{Z}$	i	$\mathbf{I}_{eff}$	$p(t)$	P	Q	S
R	R	$\frac{V\sqrt{2}}{R}\cos\omega t$	$\frac{V}{R}\angle 0^\circ$	$\frac{V^2}{R}(1+\cos 2\omega t)$	$\frac{V^2}{R}$	0	$\frac{V^2}{R}$
L	$jL\omega$	$\frac{V\sqrt{2}}{L\omega}\cos(\omega t - 90^\circ)$	$\frac{V}{L\omega}\angle -90^\circ$	$\frac{V^2}{L\omega}\sin 2\omega t$	0	$\frac{V^2}{L\omega}$	$\frac{V^2}{L\omega}$
C	$\frac{-j}{C\omega}$	$V\sqrt{2}C\omega\cos(\omega t + 90^\circ)$	$VC\omega\angle 90^\circ$	$-V^2C\omega\sin 2\omega t$	0	$-V^2C\omega$	$V^2C\omega$

Solution: Let the effective values of voltage and current be V and I, respectively.

$$p_R(t) = vi_R = (V/\sqrt{2})\cos\omega t(I/\sqrt{2})\cos\omega t = 2VI\cos^2\omega t$$
$$= VI(1+\cos 2\omega t) = RI^2(1+\cos 2\omega t) = \frac{V^2}{R}(1+\cos 2\omega t)$$

Thus,
$$P_R = \frac{V^2}{R} = RI^2, \quad Q = 0$$

The instantaneous power entering a resistor varies sinusoidally between zero and $2RI^2$. $v(t)$ and $p_R(t)$ are plotted in Figures 6-4 and 6-5, respectively.

Example 6.6 Find the ac power entering an inductor L.

Solution:

$$p_L(t) = vi_L = (V/\sqrt{2})\cos\omega t(I/\sqrt{2})\cos(\omega t - 90°) = 2VI\cos\omega t\sin\omega t$$
$$= VI\sin 2\omega t = L\omega I^2 \sin 2\omega t = \frac{V^2}{L\omega}\sin 2\omega t$$

Thus,
$$P_L = 0, \quad Q = VI = \frac{V^2}{L\omega} = L\omega I^2$$

The instantaneous power entering an inductor varies sinusoidally between $-Q$ and Q, with twice the frequency of the source, and an average value of zero. See Figure 6-6.

If an inductor and a capacitor are fed in parallel by the same ac voltage source or in series by the same current source, the power entering the capacitor is 180° out of phase with the power entering the inductor. This is explicitly reflected in the opposite signs of reactive power Q for the inductor and capacitor. In such cases, the inductor and the capacitor will exchange some energy with each other, bypassing the ac source. This reduces the reactive power delivered by the source to the LC combination and consequently improves the power factor.

Complex Power

The two components P and Q of power play different roles and may not be added together. However, they may conveniently be brought together

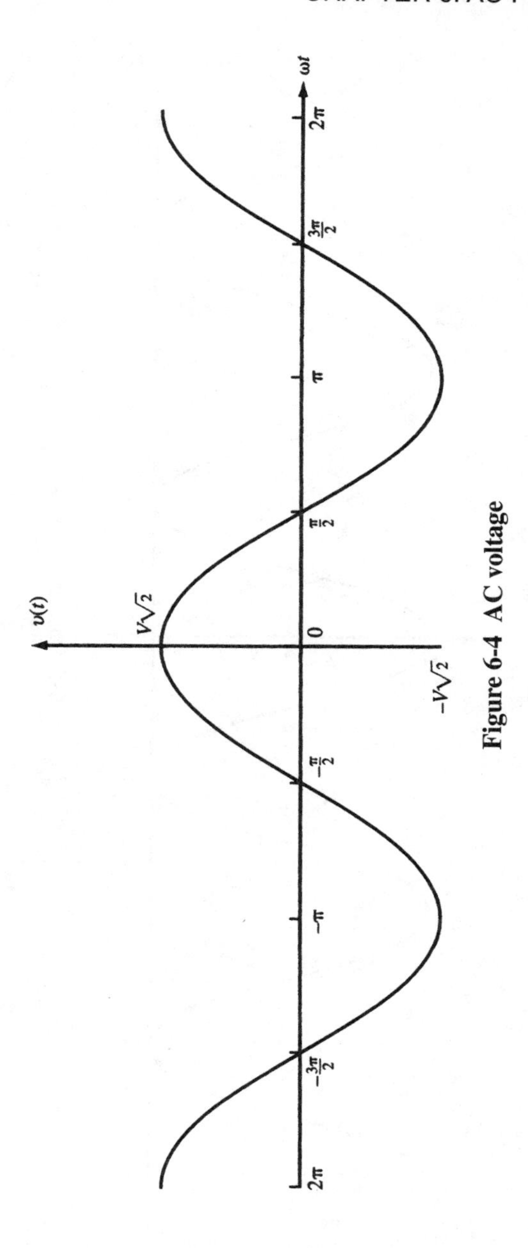

Figure 6-4 AC voltage

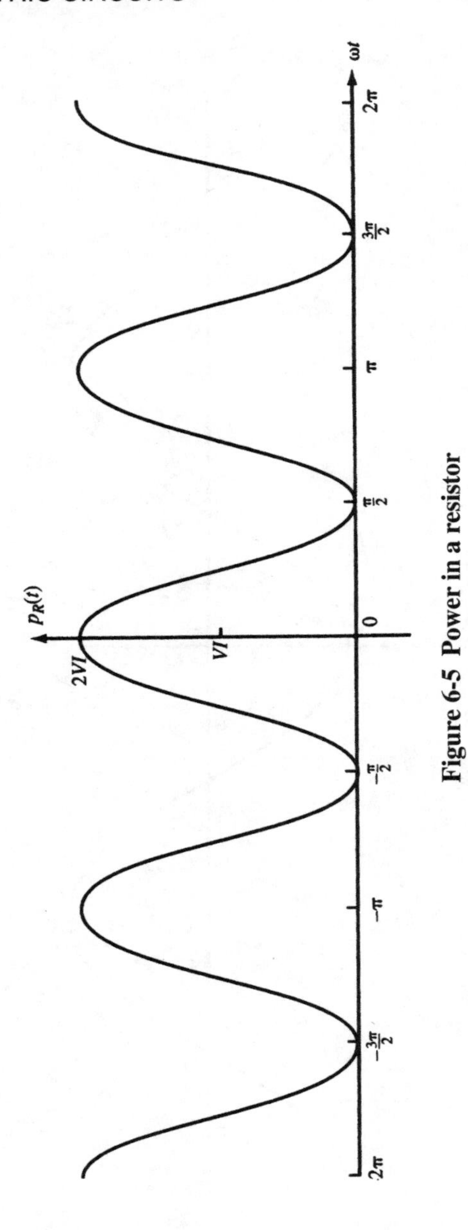

Figure 6-5 Power in a resistor

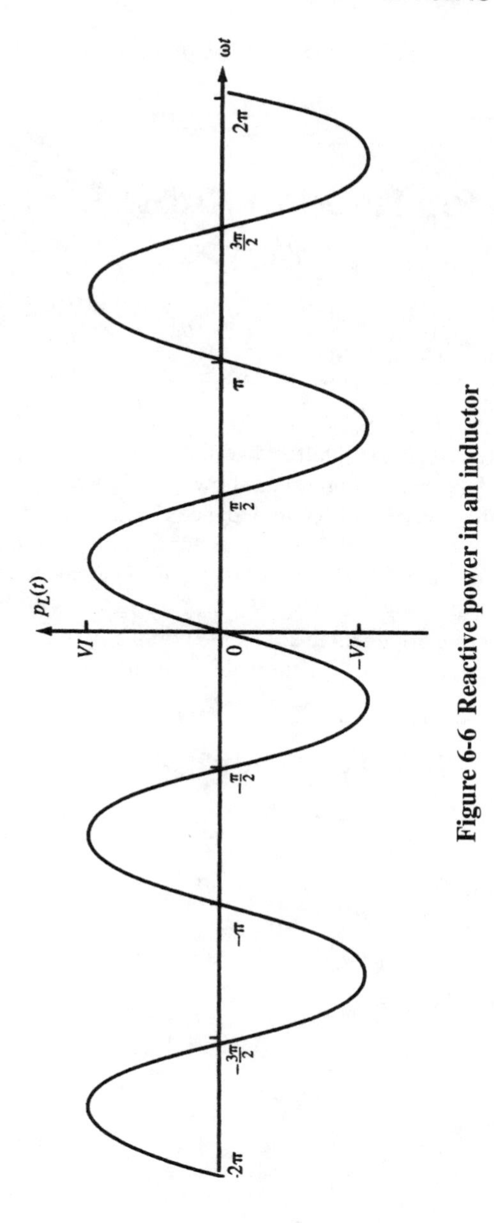

Figure 6-6 Reactive power in an inductor

in the form of a vector quantity called *complex power* **S** and defined by **S** $= P + jQ$.

You Need to Know

The magnitude $|\mathbf{S}| = \sqrt{P^2 + Q^2} = V_{eff} I_{eff}$ is called the *apparent power S* and is expressed in units of volt-amperes (VA).

The three scalar quantities S, P, and Q may be represented geometrically as the hypotenuse, horizontal, and vertical legs, respectively, of a right triangle (called a *power triangle*) as shown in Figure 6-7(a). The power triangle is simply the triangle of impedance **Z** scaled by the factor I_{eff}^2 as shown in Figure 6-7(b). Power triangles for an inductive load and a capacitive load are shown in Figures 6-7(c) and (d), respectively.

It can be easily proved that $\mathbf{S} = \mathbf{V}_{eff}\mathbf{I}_{eff}^*$, where $\mathbf{V}_{eff}$ is the complex amplitude of effective voltage and $\mathbf{I}_{eff}^*$ is the complex conjugate of the amplitude of effective current. An equivalent formula is $\mathbf{S} = \mathbf{I}_{eff}^2\mathbf{Z}$.
In summary,

Complex Power: $\mathbf{S} = \mathbf{V}_{eff}\mathbf{I}_{eff}^* = P + jQ = \mathbf{I}_{eff}^2\mathbf{Z}$

Real Power: $P = \text{Re}[\mathbf{S}] = V_{eff} I_{eff} \cos\theta$

Reactive Power: $Q = \text{Im}[\mathbf{S}] = V_{eff} I_{eff} \sin\theta$

Apparent Power: $S = V_{eff} I_{eff}$

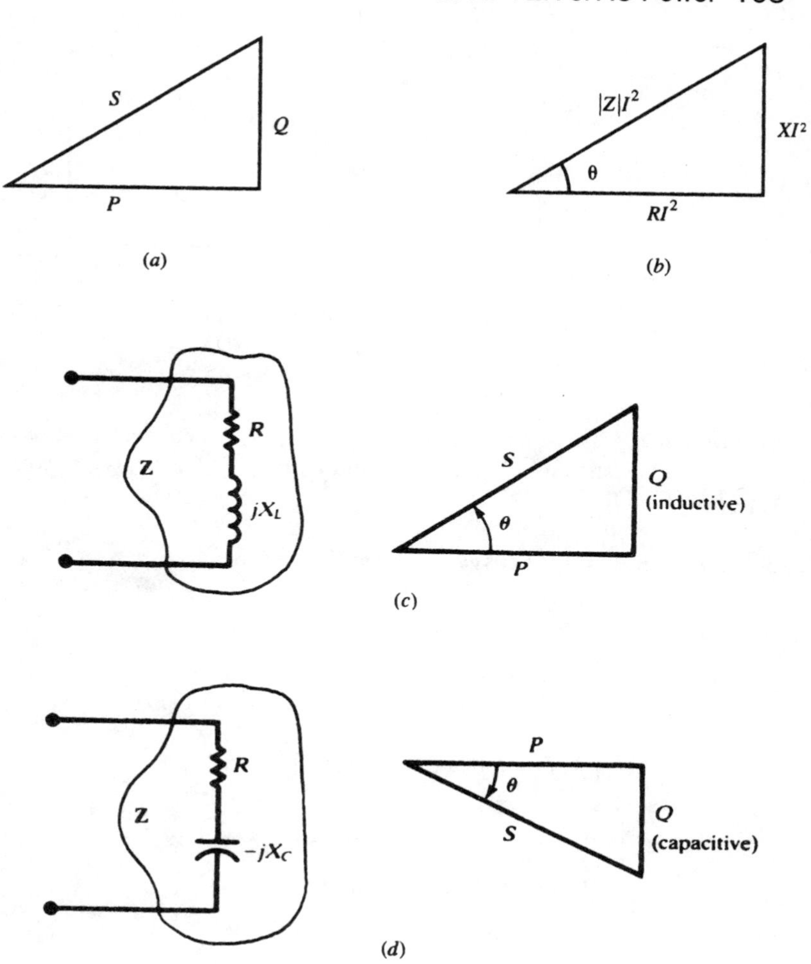

Figure 6-7 Complex power triangle examples

Figure 6-8 Complex power example

Example 6.7 A sinusoidal voltage with $V_{eff} = 10$ V is connected across $\mathbf{Z}_1 = 1 + j\Omega$ as shown in Figure 6-8. Find i_1, $\mathbf{I}_{1,eff}$, $p_1(t)$, P_1, Q_1, power factor pf_1, and $\mathbf{S}_1$.

Solution: Let $v = 10\sqrt{2}\cos\omega t$ V. Using Figure 6-8,

$$\mathbf{Z}_1 = \sqrt{2}\angle 45°\ \Omega$$

$$i_1 = 10\cos(\omega t - 45°)\ \text{A}$$

$$\mathbf{I}_{1,eff} = 5\sqrt{2}\angle -45°$$

$$\begin{aligned} p_1(t) &= (100\sqrt{2})\cos\omega t\cos(\omega t - 45°) \\ &= 50 + (50\sqrt{2})\cos(2\omega t - 45°)\ \text{W} \end{aligned}$$

$$P_1 = V_{eff} I_{1,eff}\cos 45° = 50\ \text{W}$$

$$Q_1 = V_{eff} I_{1,eff}\sin 45° = 50\ \text{var}$$

$$\mathbf{S}_1 = P_1 + jQ_1 = 50 + j50$$

$$S_1 = |\mathbf{S}_1| = 50\sqrt{2} = 70.7\ \text{VA}$$

$$\text{pf}_1 = \cos 45°\ \text{lagging} = 0.707\ \text{lagging}$$

The plots of $v(t)$, $i_1(t)$, and $p_1(t)$ are shown in Figure 6-9.

Example 6.8 A certain passive network has equivalent impedance $\mathbf{Z} = 3 + j4\Omega$ and an applied voltage $v = 42.5\cos(1000t + 30°)$V. Give complete power information.

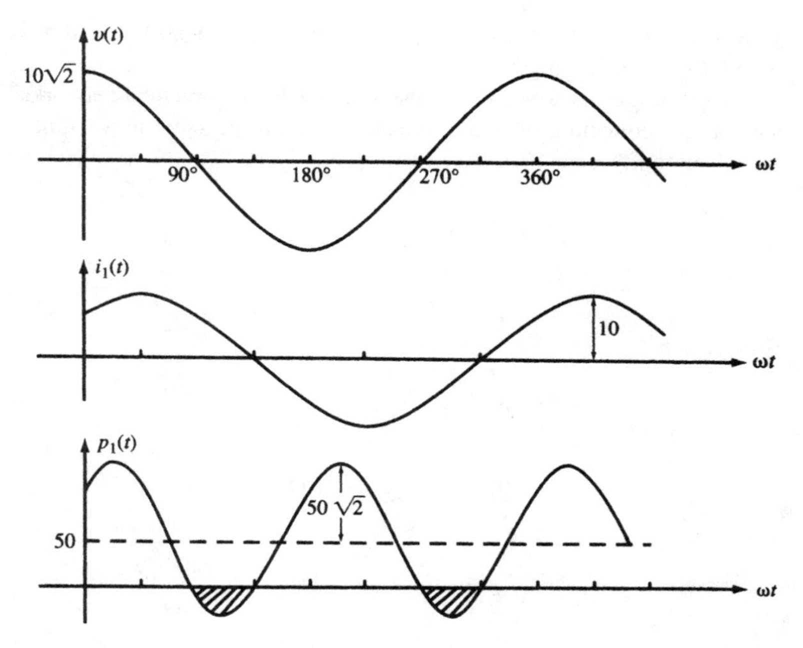

Figure 6-9 Plots of voltage, current, and power quantities

Solution:

$$\mathbf{V}_{eff} = \frac{42.5}{\sqrt{2}} \angle 30° \text{ V}$$

$$\mathbf{I}_{eff} = \frac{\mathbf{V}_{eff}}{\mathbf{Z}} = \frac{(42.5/\sqrt{2})\angle 30°}{5\angle 53.13°} = \frac{8.5}{\sqrt{2}} \angle -23.13° \text{ A}$$

$$\mathbf{S} = \mathbf{V}_{eff}\mathbf{I}^{*}_{eff} = 180.6\angle 53.13° = 108.4 + j144.5$$

Hence, $P = 108.4$ W, $Q = 144.5$ var (inductive), $S = 180.6$ VA, and pf = cos53.13° = 0.6 lagging.

The complex power **S** is also useful in analyzing practical networks, such as the collection of households drawing on the same power lines. Referring to Figure 6-10,

$$\begin{aligned}\mathbf{S}_T &= \mathbf{V}_{eff}\mathbf{I}^*_{eff} = \mathbf{V}_{eff}(\mathbf{I}^*_{1,eff} + \mathbf{I}^*_{2,eff} + \ldots + \mathbf{I}^*_{n,eff}) \\ &= \mathbf{S}_1 + \mathbf{S}_2 + \ldots\mathbf{S}_n\end{aligned}$$

from which

$$P_T = P_1 + P_2 + \ldots + P_n$$

$$Q_T = Q_1 + Q_2 + \ldots + Q_n$$

$$S_T = \sqrt{P_T^2 + Q_T^2}$$

$$\text{pf}_T = \frac{P_T}{S_T}$$

These results (which also hold for series-connected networks) imply that the power triangle for the network may be obtained by joining the power triangles for the branches vertex to vertex. In the example shown in Figure 6-11, $n = 3$, with branches 1 and 3 assumed inductive and branch

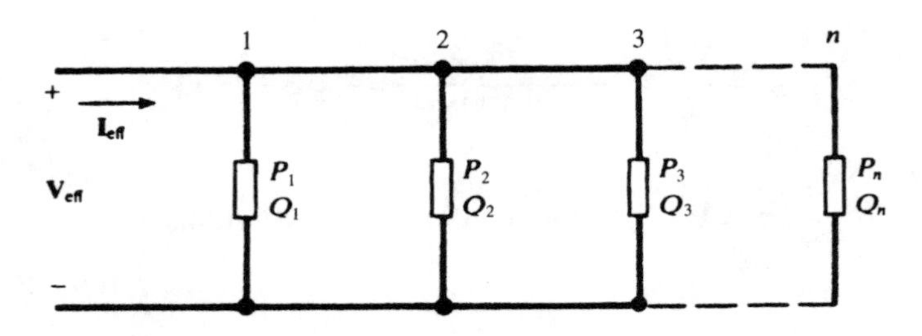

Figure 6-10 Parallel-connected network

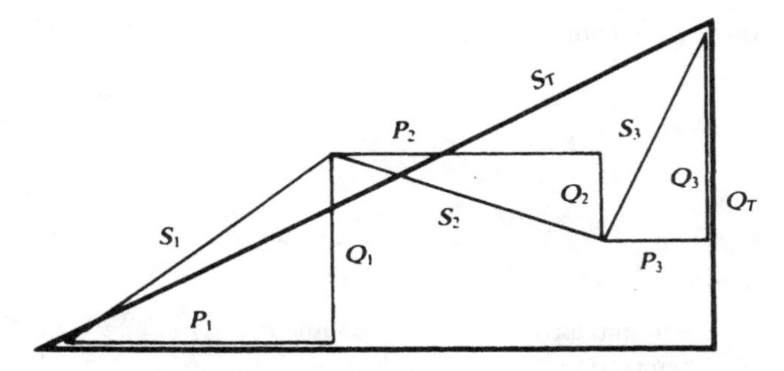

Figure 6-11 Power triangle for parallel-connected network

2 capacitive. In such diagrams, some of the triangles may degenerate into straight-line segments if the corresponding R or X is zero.

While metering and billing practices vary among the utilities, the large consumers will always find it advantageous to reduce the quadrature component of their power triangle. This is called *power factor correction*. Industrial systems generally have an overall inductive component because of the large number of motors. Each individual load tends to be either pure resistance, with unity power factor, or resistance and inductive reactance, with a lagging power factor. All of the loads are parallel-connected and the equivalent impedance results in a lagging current and a corresponding inductive quadrature power Q. To improve the power factor, capacitors are connected to the system either on the primary or secondary side of the main transformer, such that the combination of the plant load and the capacitor banks presents a load to the serving utility that is nearer to unity power factor.

Example 6.9 How much capacitive Q must be provided by the capacitor bank in Figure 6-12 to improve the power factor to 0.95 lagging?

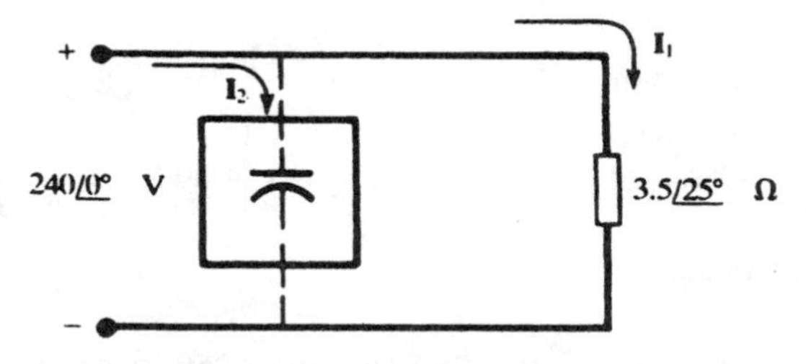

Figure 6-12 Power factor improvement example

Solution: Before addition of the capacitor bank, pf = cos25° = 0.906 lagging, and

$$\mathbf{I}_1 = \frac{240\angle 0°}{3.5\angle 25°} = 68.6\angle -25° \text{ A}$$

$$\mathbf{S} = \mathbf{V}_{eff}\mathbf{I}^*_{eff} = \left(\frac{240\angle 0°}{\sqrt{2}}\right)\left(\frac{68.6\angle +25°}{\sqrt{2}}\right) = 8232\angle 25° = 7461 + j3479$$

After improvement, the triangle has the same P, but its angle is $\cos^{-1}0.95 = 18.19°$. Then (see Figure 6-13),

$$\frac{3479 - Q_c}{7461} = \tan 18.19° \text{ or } Q_c = 1027 \text{ var (capacitive)}$$

The new value of apparent power is $S' = 7854$ VA, as compared to the original $S = 8232$ VA. The decrease, 378 VA, amounts to 4.6 percent.

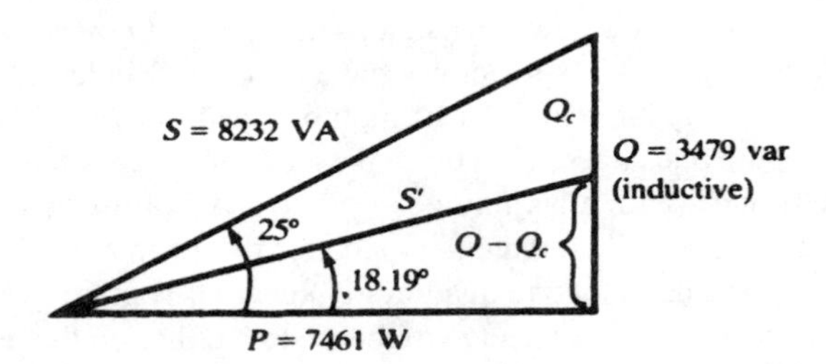

Figure 6-13 Power triangle for Example 6.9

Example 6.10 A load of $P = 1000$ kW with a pf = 0.5 lagging is fed by a 5-kV source. A capacitor is added in parallel such that the power factor is improved to 0.8. Find the reduction in current drawn from the generator.

Solution: Before improvement:

$$P = 1000 \text{ kW}, \cos\theta = 0.5, S = P/\cos\theta = 2000 \text{ kVA}, I = 400 \text{ A}$$

After improvement:

$$P = 1000 \text{ kW}, \cos\theta = 0.8, S = P/\cos\theta = 1250 \text{ kVA}, I = 250 \text{ A}$$

The average power delivered to a load $\mathbf{Z}_1$ from a sinusoidal signal generator with open circuit voltage $\mathbf{V}_g$ and internal impedance $\mathbf{Z}_g = R + jX$ is maximum when $\mathbf{Z}_1$ is equal to the complex conjugate of $\mathbf{Z}_g$ so that $\mathbf{Z}_1 = R - jX$. This is the condition for *maximum power transfer* when complex impedances are concerned.

Example 6.11 A generator, with $V_g = 100$ V (rms) and $\mathbf{Z}_g = 1 + j\Omega$, feeds a load $\mathbf{Z}_1 = 2\Omega$ (Figure 6-14). (*a*) Find the average power P_{Z1} (absorbed by $\mathbf{Z}_1$), the power P_g (dissipated by $\mathbf{Z}_g$) and P_T (provided by the generator). (*b*) Compute the value of a second load $\mathbf{Z}_2$ such that, when in parallel with $\mathbf{Z}_1$, the equivalent impedance is $\mathbf{Z} = \mathbf{Z}_1 \| \mathbf{Z}_2 = \mathbf{Z}_g^*$. (*c*) Connect in parallel $\mathbf{Z}_2$ found in (*b*) with $\mathbf{Z}_1$ and then find the powers P_Z, P_{Z1}, P_{Z2} (absorbed by $\mathbf{Z}$, $\mathbf{Z}_1$, and $\mathbf{Z}_2$, respectively), P_g (dissipated in $\mathbf{Z}_g$) and P_T (provided by the generator).

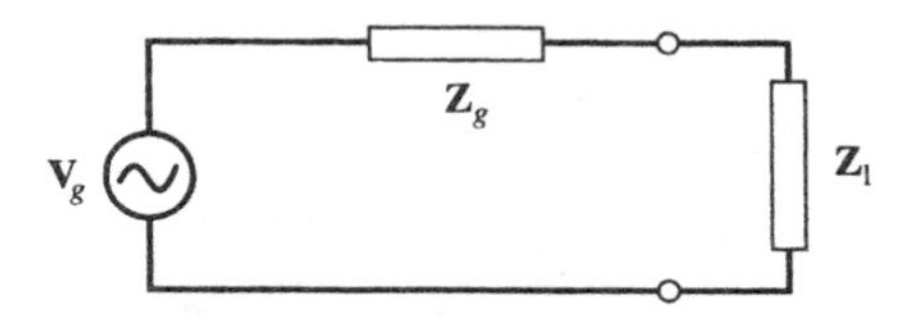

Figure 6-14 Maximum power transfer example circuit

Solution:

(*a*) $|\mathbf{Z}_1 + \mathbf{Z}_g| = |2 + 1 + j| = \sqrt{10}$.

Thus, $\mathbf{I} = \mathbf{V}_g / (\mathbf{Z}_1 + \mathbf{Z}_g) = 100/(2 + 1 + j)$ and $|\mathbf{I}| = 10\sqrt{10}$ A. The required powers are

$$P_{Z1} = \mathrm{Re}[\mathbf{Z}_1] \times |\mathbf{I}|^2 = 2(10\sqrt{10})^2 = 2000 \text{ W}$$

$$P_g = \mathrm{Re}[\mathbf{Z}_g] \times |\mathbf{I}|^2 = 1(10\sqrt{10})^2 = 1000 \text{ W}$$

$$P_T = P_{Z1} + P_g = 2000 + 1000 = 3000 \text{ W}$$

(*b*) Let $\mathbf{Z}_2 = a + jb$. To find a and b, we set $\mathbf{Z}_1 \| \mathbf{Z}_2 = 1 - j$. Then,

$$\frac{\mathbf{Z}_1\mathbf{Z}_2}{\mathbf{Z}_1 + \mathbf{Z}_2} = \frac{2(a + jb)}{2 + a + jb} = 1 - j$$

from which $a - b - 2 = 0$ and $a + b + 2 = 0$. Solving these simultaneous equations, $a = 0$ and $b = -2$. Substituting into the equation above, $\mathbf{Z}_2 = -j2$.

(*c*) $\mathbf{Z} = \mathbf{Z}_1 \| \mathbf{Z}_2 = 1 - j$ and $\mathbf{Z} + \mathbf{Z}_g = 1 - j + 1 + j = 2$. Then,
$\mathbf{I} = \mathbf{V}_g / (\mathbf{Z} + \mathbf{Z}_g) = 100/2 = 50$ A, and so

$$P_Z = \mathrm{Re}[\mathbf{Z}] \times |\mathbf{I}|^2 = 1(50)^2 = 2500 \text{ W}$$

$$P_g = \mathrm{Re}[\mathbf{Z}_g] \times |\mathbf{I}|^2 = 1(50)^2 = 2500 \text{ W}$$

To find P_{Z1} and P_{Z2}, we first find $\mathbf{V}_Z$ across $\mathbf{Z}$: $\mathbf{V}_Z = \mathbf{IZ} = 50(1 - j)$. Then $\mathbf{I}_{Z1} = \mathbf{V}_Z / \mathbf{Z}_1 = 50(1 - j)/2 = (25\sqrt{2})\angle -45°$ and

$$P_{Z1} = \mathrm{Re}[\mathbf{Z}_1] \times |\mathbf{I}_{Z1}|^2 = 2(25\sqrt{2})^2 = 2500 \text{ W}$$

$$P_{Z2} = 0 \text{ W}$$

$$P_T + P_{Z1} + P_g = 5000 \text{ W}$$

Important Things to Remember

- ✔ In a circuit analysis, the magnitude of the effective voltage V_{eff} is the rms voltage; likewise for the current.
- ✔ The ratio of P_{avg} to $V_{eff} I_{eff}$ is called the *power factor* pf and is equal to $\cos\theta$, where θ is the angle between the voltage and current.
- ✔ An inductive load has a *lagging* power factor while a capacitive load has a *leading* power factor.
- ✔ The reactive power Q in a purely resistive load is zero.
- ✔ In a network, the *complex power* **S** is defined by $\mathbf{S} = P + jQ$.
- ✔ The metering and billing practices of utilities provides economic incentive for industries to optimize the power factors of their networks (design for pf to approach unity).
- ✔ Maximum power is transferred to a load whose impedance is the complex conjugate of the source impedance.

Chapter 7
FREQUENCY RESPONSE, FILTERS, AND RESONANCE

IN THIS CHAPTER:

- ✔ *Frequency Response*
- ✔ *Low-, High-, and Band-Pass Filters*

Frequency Response

The response of linear circuits to a sinusoidal input is also a sinusoid, with the same frequency but possibly a different amplitude and phase angle. This response is a function of the frequency. We have already seen that a sinusoid can be represented by a phasor that shows its magnitude and phase. The *frequency response* is defined as the ratio of the output phasor to the input phasor. It is a function of ω and is given by

$$\mathbf{H}(\omega) = \text{Re}[\mathbf{H}] + j\,\text{Im}[\mathbf{H}] = |\mathbf{H}|\,e^{j\theta}$$

where Re[**H**] and Im[**H**] are the real and imaginary parts of $\mathbf{H}(\omega)$ and $|\mathbf{H}|$ and θ are its magnitude and phase angle. Re[**H**], Im[**H**], $|\mathbf{H}|$ and θ are, in general, functions of ω. They are related by

$$|\mathbf{H}|^2 = |\mathbf{H}(\omega)|^2 = \mathrm{Re}^2[\mathbf{H}] + \mathrm{Im}^2[\mathbf{H}]$$

$$\theta = \angle\mathbf{H}(\omega) = \tan^{-1}\frac{\mathrm{Im}[\mathbf{H}]}{\mathrm{Re}[\mathbf{H}]}$$

The frequency response, therefore, depends on the choice of input and output variables. For example, if a current source is connected across the network of Figure 7-1, the terminal current is the input and the terminal voltage may be taken as the output. In this case, the input impedance $\mathbf{Z} = \mathbf{V}_1/\mathbf{I}_1$ constitutes the frequency response. Conversely, if a voltage source is applied to the input and the terminal current is measured, the input admittance $\mathbf{Y} = \mathbf{I}_1/\mathbf{V}_1$ represents the frequency response.

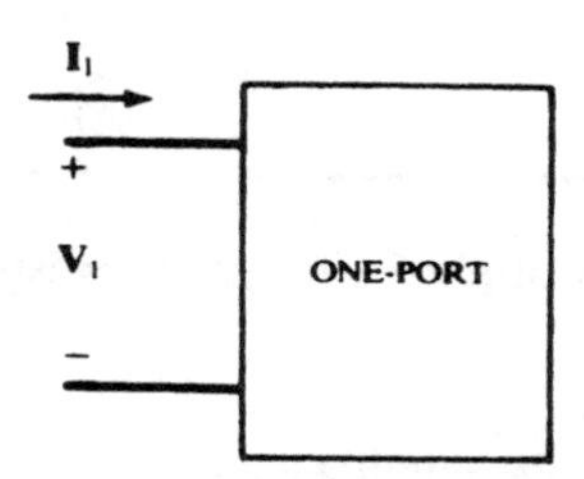

Figure 7-1 Frequency response for a one-port network

For the two-port network of Figure 7-2, the following frequency responses are defined:

Input impedance: $\mathbf{Z}_{in}(\omega) = \mathbf{V}_1/\mathbf{I}_1$

Input admittance: $\mathbf{Y}_{in}(\omega) = 1/\mathbf{Z}_{in}(\omega) = \mathbf{I}_1/\mathbf{V}_1$

Voltage transfer ratio: $\mathbf{H}_v(\omega) = \mathbf{V}_2/\mathbf{V}_1$

Current transfer ratio: $\mathbf{H}_i(\omega) = \mathbf{I}_2/\mathbf{I}_1$

Transfer impedances: $\mathbf{V}_2/\mathbf{I}_1$ and $\mathbf{V}_1/\mathbf{I}_2$

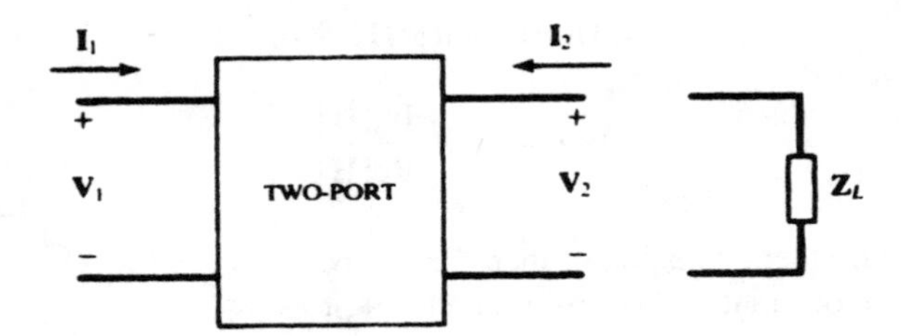

Figure 7-2 Frequency response for a two-port network

Example 7.1 Find the frequency response $\mathbf{V}_2/\mathbf{V}_1$ for the two-port network shown in Figure 7-3.

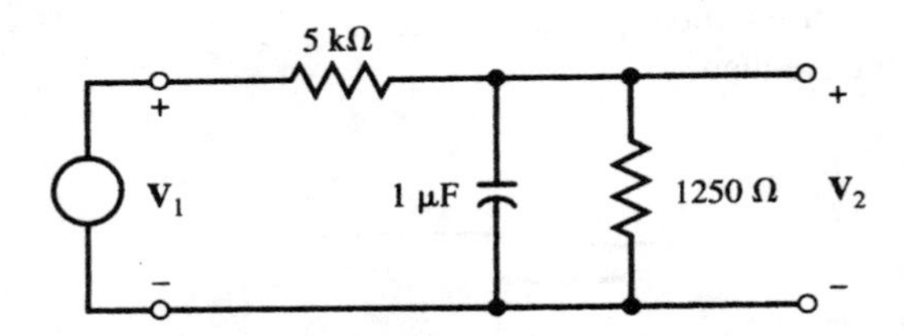

Figure 7-3 Example 7.1 frequency response network

Solution: Let $\mathbf{Y}_{RC}$ be the admittance of the parallel RC combination. Then, $\mathbf{Y}_{RC} = 10^{-6} j\omega + 1/1250$. $\mathbf{V}_2/\mathbf{V}_1$ is obtained by dividing $\mathbf{V}_1$ between $\mathbf{Z}_{RC}$ and the 5-kΩ resistor.

$$\mathbf{H}(\omega) = \frac{\mathbf{V}_2}{\mathbf{V}_1} = \frac{\mathbf{Z}_{RC}}{\mathbf{Z}_{RC} + 5000} = \frac{1}{1 + 5000\mathbf{Y}_{RC}} = \frac{1}{5(1 + 10^{-3} j\omega)}$$

$$|\mathbf{H}| = \frac{1}{5\sqrt{1 + 10^{-6}\omega^2}}, \ \theta = -\tan^{-1}(10^{-3}\omega)$$

Low-, High-, and Band-Pass Filters

A resistive voltage divider under a no-load condition is shown in Figure 7-4, with the standard two-port voltages and currents.

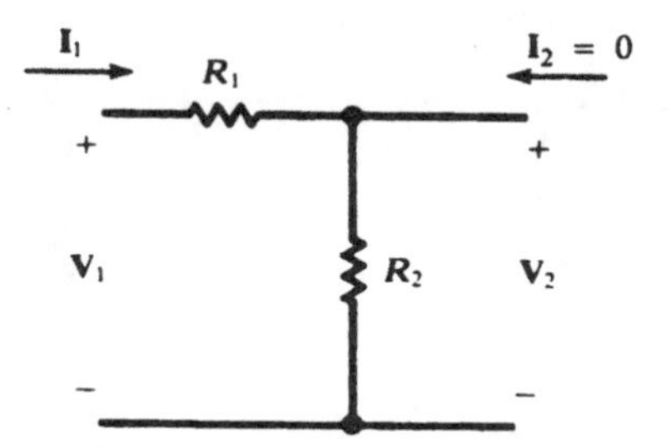

Figure 7-4 Resistive network

The voltage transfer function and input impedance are

$$\mathbf{H}_{v\infty}(\omega) = \frac{R_2}{R_1 + R_2} \quad \text{and} \quad \mathbf{H}_{z\infty}(\omega) = R_1 + R_2$$

The ∞ in the subscripts indicates no-load conditions. Both $\mathbf{H}_{v\infty}$ and $\mathbf{H}_{z\infty}$ are real constants, independent of frequency, since no reactive elements are present. If the network contains either an inductance or a capacitance, then $\mathbf{H}_{v\infty}$ and $\mathbf{H}_{z\infty}$ will be complex and will vary with frequency.

If $|\mathbf{H}_{v\infty}|$ decreases as frequency increases, the performance is called *high-frequency roll-off* and the circuit is a *low-pass network or low-pass filter.* If $|\mathbf{H}_{v\infty}|$ increases as frequency increases, the performance is called *low-frequency roll-off* and the circuit is a *high-pass network* or *high-pass filter.*

Four two-element circuits are shown in Figure 7-5, two high-pass and two low-pass.

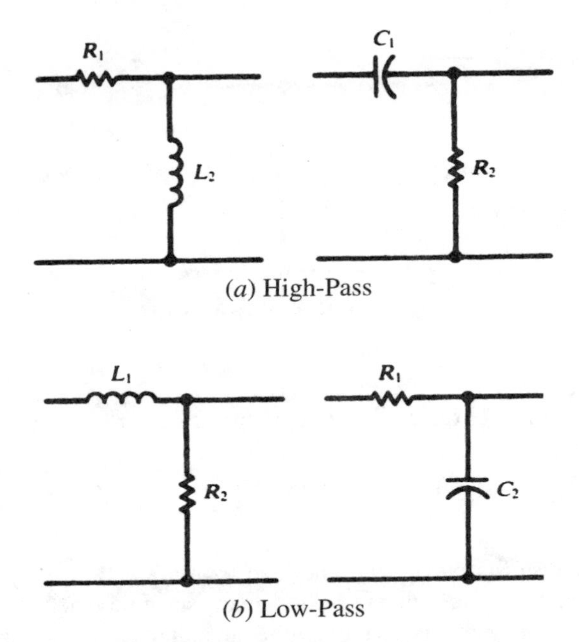

Figure 7-5 High- and low-pass filter circuits

The *RL* high-pass circuit shown in Figure 7-6 is open-circuited under no-load. The input impedance frequency response is determined by plotting the magnitude and phase angle of

$$\mathbf{H}_{z\infty}(\omega) = R_1 + j\omega L_2 \equiv |\mathbf{H}_z| \angle\theta_H$$

or, normalizing and writing $\omega_x \equiv R_1/L_2$,

$$\frac{\mathbf{H}_{z\infty}(\omega)}{R_1} = 1 + j(\omega/\omega_x) = \sqrt{1+(\omega/\omega_x)^2}\angle\tan^{-1}(\omega/\omega_x)$$

Five values of ω provide sufficient data to plot $|\mathbf{H}_z|/R_1$ and θ_H, as shown in Figure 7-7. The magnitude approaches infinity with increasing frequency, and so, at very high frequencies, the network current $\mathbf{I}_1$ will be zero.

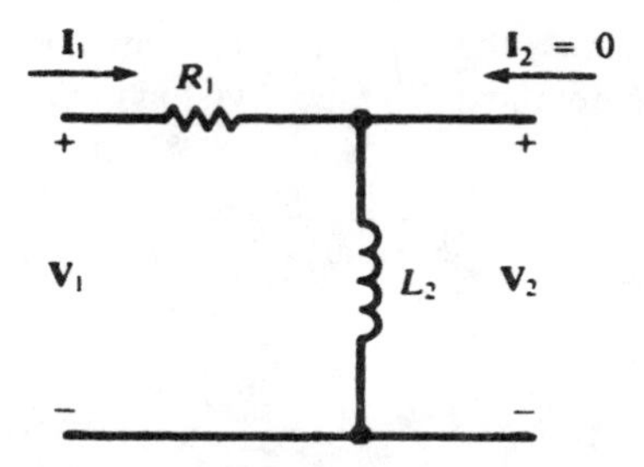

Figure 7-6 High-pass circuit

ω	$\lvert \mathbf{H}_z \rvert / R_1$	θ_H
0	1	0°
$0.5\omega_x$	$0.5\sqrt{5}$	26.6°
ω_x	$\sqrt{2}$	45°
$2\omega_x$	$\sqrt{5}$	63.4°
∞	∞	90°

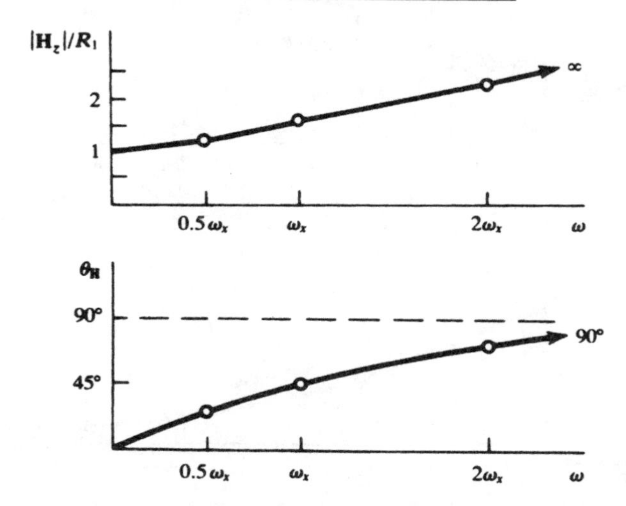

Figure 7-7 High-pass input impedance of the *RL* circuit

In a similar manner, the frequency response of the output-to-input voltage ratio can be obtained. Voltage division under no-load gives

$$\mathbf{H}_{v\infty}(\omega) = \frac{j\omega L_2}{R_1 + j\omega L_2} = \frac{1}{1 - j(\omega_x / \omega)}$$

so that

$$|\mathbf{H}_v| = \frac{1}{\sqrt{1 + (\omega_x / \omega)^2}} \text{ and } \theta_H = \angle \tan^{-1}(\omega_x / \omega)$$

The magnitude and angle are plotted in Figure 7-8. This transfer function approaches unity at high frequency, where the output voltage is the same as the input. Hence, the description "low-frequency roll-off" and the name "high-pass."

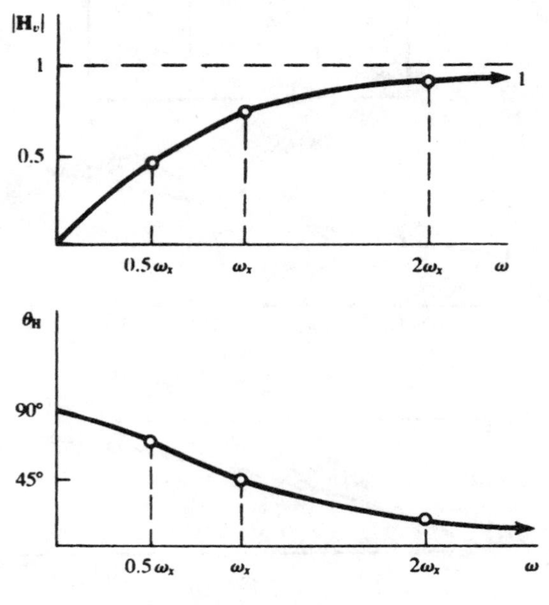

Figure 7-8 High-pass voltage ratio

A transfer impedance of the *RL* high-pass circuit under no-load is

$$\mathbf{H}_{\infty}(\omega) = \frac{\mathbf{V}_2}{\mathbf{I}_1} = j\omega L_2 \text{ or } \frac{\mathbf{H}_{\infty}(\omega)}{R_1} = j\frac{\omega}{\omega_x}$$

The angle is constant at 90°; the graph of the magnitude versus ω is a straight line, similar to a reactance plot of ωL versus ω. See Figure 7-9.

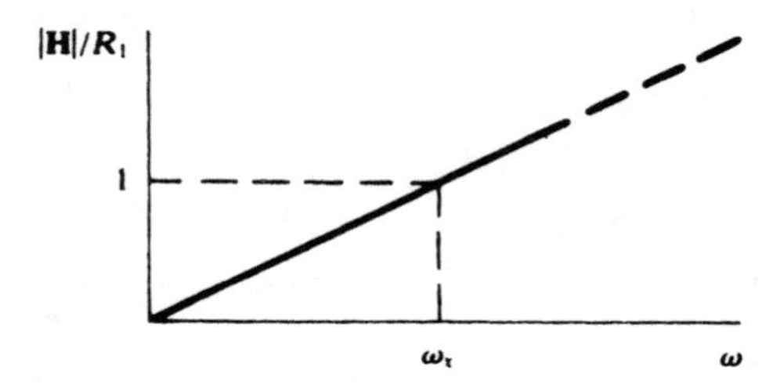

Figure 7-9 High-pass transfer impedance plot

Interchanging the positions of *R* and *L* results in a low-pass network with high-frequency roll-off (Figure 7-10).

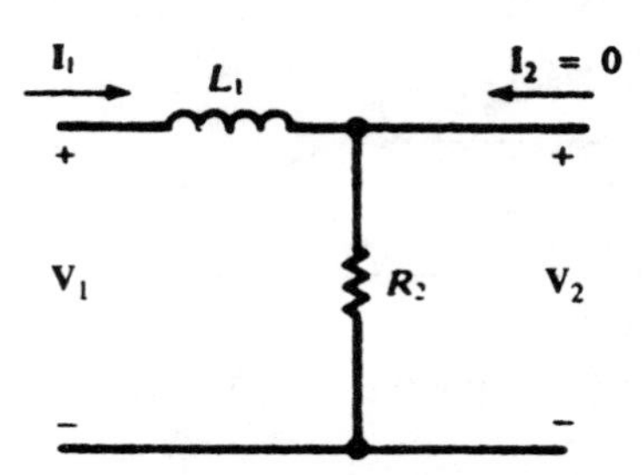

Figure 7-10 Low-pass circuit

For the open-circuit condition

$$\mathbf{H}_{v\infty}(\omega) = \frac{R_2}{R_2 + j\omega L_1} = \frac{1}{1 + j(\omega / \omega_x)}$$

with $\omega_x \equiv R_2/L_1$; that is,

$$|\mathbf{H}_v| = \frac{1}{\sqrt{1+(\omega/\omega_x)^2}} \quad \text{or} \quad \theta_H = \angle \tan^{-1}(-\omega/\omega_x)$$

The magnitude and phase plots are shown in Figure 7-11. The voltage transfer function $\mathbf{H}_{v\infty}$ approaches zero at high frequencies and unity at $\omega = 0$. Hence, the name "low-pass."

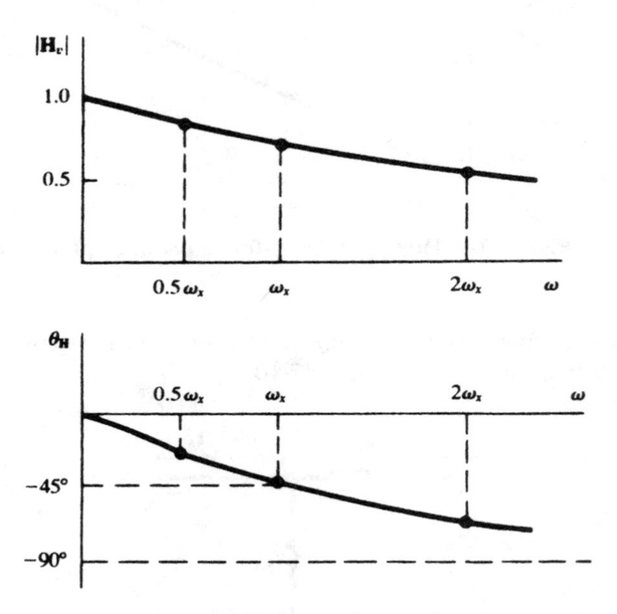

Figure 7-11 $\mathbf{H}_v$ for the low-pass network of Figure 7-10

Other network functions for other circuits can be obtained in similar fashions.

Example 7.2 Obtain the voltage transfer function $\mathbf{H}_{v\infty}$ for the open circuit shown in Fig. 7-12. At what frequency, in Hz, does $|\mathbf{H}_v| = 1/\sqrt{2}$ if (*a*) $C_2 = 10$ nF, (*b*) $C_2 = 1$ nF?

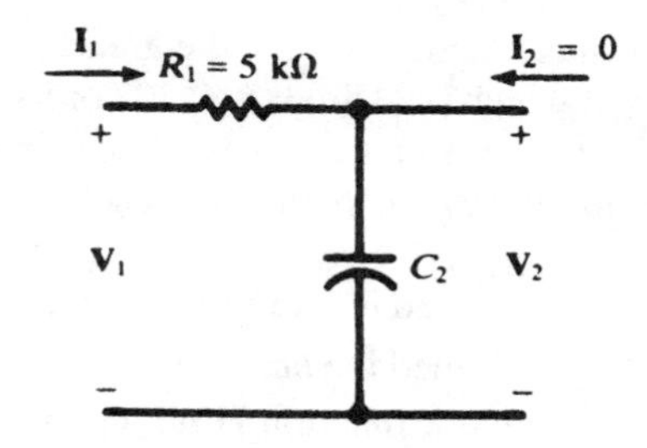

Figure 7-12 Filter circuit for Example 7.2

Solution:

$$\mathbf{H}_{v\infty}(\omega) = \frac{1/j\omega C_2}{R_1 + 1/j\omega C_2} = \frac{1}{1 + j(\omega/\omega_x)}$$

where

$$\omega_x \equiv \frac{1}{R_1 C_2} = \frac{2\times 10^{-4}}{C_2} \text{ (rad/s)}$$

and

$$|\mathbf{H}_v| = \frac{1}{\sqrt{1+(\omega/\omega_x)^2}}$$

(a) $C_2 = 10$ nF $\Rightarrow |\mathbf{H}_v| = 1/\sqrt{2}$ when

$$\omega = \omega_x = \frac{2\times 10^{-4}}{10\times 10^{-9}} = 2\times 10^4 \text{(rad/s)}$$

or when $f = 2 \times 10^4/2\pi = 3.18$ kHz

(b) $C_2 = 10$ nF $\Rightarrow |\mathbf{H}_v| = 1/\sqrt{2}$ when

$$\omega = \omega_x = \frac{2\times 10^{-4}}{1\times 10^{-9}} = 2\times 10^5 \text{(rad/s)}$$

or when $f = 2 \times 10^5/2\pi = 31.8$ kHz

Comparing (*a*) and (*b*), it is seen that the greater the value of C_2, the lower the frequency at which $|\mathbf{H}_v|$ drops to 0.707 of its peak value of unity. Consequently, any stray shunting capacitance, in parallel with C_2, serves to reduce the frequency where the frequency response drops to the 0.707 value.

The frequency ω_x calculated in Example 7.2, the frequency at which $|\mathbf{H}_v| = (1/\sqrt{2})|\mathbf{H}_v|_{max}$, is called the *half-power frequency*. Quite generally, any non-constant network function $\mathbf{H}(\omega)$ will attain its greatest absolute value at some unique frequency ω_{max}.

We shall call a frequency at which

$$|\mathbf{H}(\omega)| = 0.707 |\mathbf{H}(\omega_{max})|$$

a *half-power frequency* (or *half-power point*), whether or not this frequency actually corresponds to 50 percent power.

In most cases, $0 < \omega_{max} < \infty$, so that there are two half-power frequencies; one above and one below the peak frequency. These are called the *upper* and *lower* half-power frequencies (points), and their separation, the *bandwidth*, serves as a measure of the sharpness of the peak.

The following function is called a *bandpass function*. The frequency response of the bandpass function is

$$\mathbf{H}(\omega) = \frac{kj\omega}{b - \omega^2 + aj\omega}, |\mathbf{H}|^2 = \frac{k^2}{a^2 + (b - \omega^2)^2 / \omega^2} \tag{1}$$

The maximum of $|\mathbf{H}|$ occurs when $(b - \omega^2) = 0$ or $\omega = \sqrt{b}$, which is called the *center frequency* ω_0. At the center frequency, we have $|\mathbf{H}|_{max} = |\mathbf{H}(\omega_0)| = k/a$. The half-power frequencies are at ω_l and ω_h, where

$$|\mathbf{H}(\omega_l)|^2 = |\mathbf{H}(\omega_h)|^2 = \frac{1}{2}|\mathbf{H}(\omega_0)|^2 \tag{2}$$

By applying (1) to (2), ω_l and ω_h are found to be the roots of the following equation:

$$(b-\omega^2)^2 / \omega^2 = a^2 \tag{3}$$

Solving,

$$\omega_l = \sqrt{a^2/4+b} - a/2 \tag{4}$$

$$\omega_h = \sqrt{a^2/4+b} + a/2 \tag{5}$$

From (4) and (5), we have

$$\omega_h - \omega_l = a \text{ and } \omega_h \omega_l = b = \omega_0^2$$

The *bandwidth* β is defined by

$$\beta = \omega_h - \omega_l = a$$

The *quality factor* Q is defined by

$$Q = \omega_0 / \beta = \sqrt{b} / a$$

The quality factor measures the sharpness of the frequency response around the center frequency. When the quality factor is high, ω_l and ω_h may be approximated by $\omega_0 - \beta/2$ and $\omega_0 + \beta/2$, respectively.

Example 7.3 Consider the network function

$$\mathbf{H}(\omega) = \frac{10j\omega}{10^6 - \omega^2 + 300j\omega}$$

Find the center frequency, lower and upper half-power frequencies, the bandwidth, and the quality factor.

Solution: Since $b = \omega_0^2 = 10^6 \Rightarrow \omega_0 = 1000 \text{ rad/s}$. The lower and upper half-power frequencies are, respectively,

$$\omega_l = \sqrt{a^2/4+b} - a/2 = \sqrt{300^2/4+10^6} - 300/2 = 861.2 \text{ rad/s}$$

$$\omega_h = \sqrt{a^2/4+b} + a/2 = \sqrt{300^2/4+10^6} + 300/2 = 1161.2 \text{ rad/s}$$

The bandwidth is

$$\beta = \omega_h - \omega_l = a = 1161.2 - 861.2 = 300 \text{ rad/s}.$$

The quality factor $Q = 1000/300 = 3.3$.

As an example of a bandpass network, consider the *RLC* circuit in Figure 7-13.

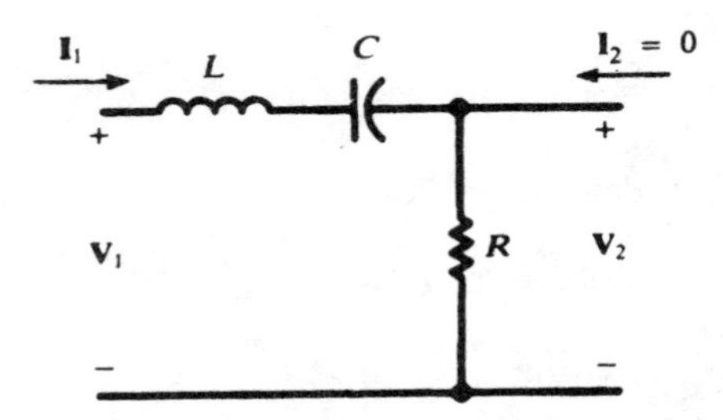

Figure 7-13 Series *RLC* circuit

Under open-circuit conditions, the *RLC* circuit has input or driving-point impedance

$$\mathbf{Z}_{in}(\omega) = R + j\left(\omega L - \frac{1}{\omega C}\right)$$

The circuit is said to be in *series resonance* when $\mathbf{Z}_{in}(\omega)$ is real (and also has minimum magnitude); that is, when

$$\omega L - \frac{1}{\omega C} = 0 \text{ or } \omega = \omega_0 \equiv \frac{1}{\sqrt{LC}}$$

By voltage division, the voltage transfer function for the circuit in Figure 7-13 is

$$\mathbf{H}_{v\infty}(\omega) = \frac{R}{\mathbf{Z}_{in}(\omega)} = R\mathbf{Y}_{in}(\omega)$$

The frequency response magnitude is plotted in Figure 7-14. Note that roll-off occurs both below and above ω_0. The quality factor for the series resonant circuit is given by $Q_0 = \omega_0 L/R$ and the bandwidth can be shown to be $\beta = \omega_h - \omega_l = R/L = \omega_0/Q_0$. These imply that the smaller the resistance R, the higher the quality factor and the smaller the bandwidth.

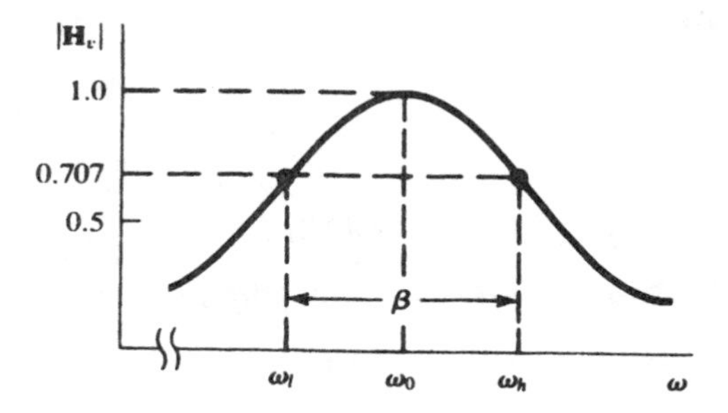

Figure 7-14 Voltage frequency response for series *RLC*

Important Things to Remember

- ✔ In general, the *frequency response* for a network is a complex function representing the ratio of output to input.
- ✔ If the frequency response increases with increasing frequency, it is said to be a *high-pass* response.
- ✔ If the frequency response decreases with increasing frequency, it is said to be a *low-pass* response.
- ✔ A bandpass response has roll-off on both sides of the resonant frequency ω_0.
- ✔ In general, for a bandpass circuit, the higher the quality factor, the narrower the bandwidth.

Appendix
COMPLEX NUMBER SYSTEM

IN THE APPENDIX:

- ✔ *Complex Numbers*
- ✔ *Sum and Difference of Complex Numbers*
- ✔ *Multiplication of Complex Numbers*
- ✔ *Division of Complex Numbers*
- ✔ *Conjugate of a Complex Number*

Complex Numbers

A *complex number* $\mathbf{z}$ is a number of the form $x + jy$, where x and y are real numbers and $j = \sqrt{-1}$. We write $x = \text{Re}\mathbf{z}$, the *real part of* $\mathbf{z}$; $y = \text{Im}\mathbf{z}$, the *imaginary part of* $\mathbf{z}$. Two complex numbers are equal if and only if their real parts are equal and their imaginary parts are equal.

A pair of orthogonal axes, with the horizontal axis displaying Re $\mathbf{z}$ and the vertical axis j Im $\mathbf{z}$, determine a complex plane in which each complex number is a unique point. Refer to Figure A-1, on which 6 complex numbers are shown. Equivalently, each complex number is represented by a unique vector from the origin of the complex plane, as illustrated for the complex number $\mathbf{z}_6$ in Figure A-1.

In addition to the definition of j above, it may be viewed as an operator that rotates any complex number (vector) $\mathbf{A}$ 90° in the counter-

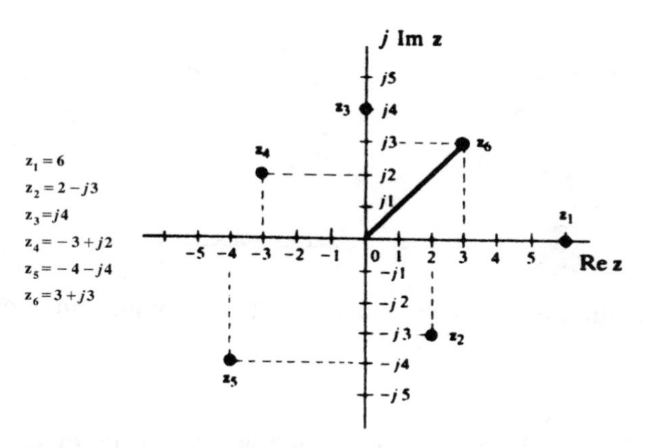

Figure A-1 Complex numbers in a complex plane

clockwise direction. The case where **A** is a pure real number x is illustrated in Figure A-2. The rotation sends **A** into jx, on the positive imaginary axis. Continuing, j^2 advances **A** 180°; j^3,270°; and j^4, 360°. Also shown in Figure A-2 is a complex number **B** in the first quadrant, at angle θ. Note that j**B** is in the second quadrant, at angle $\theta + 90°$.

The form "$x + jy$" is said to be the *rectangular form* for the complex number.

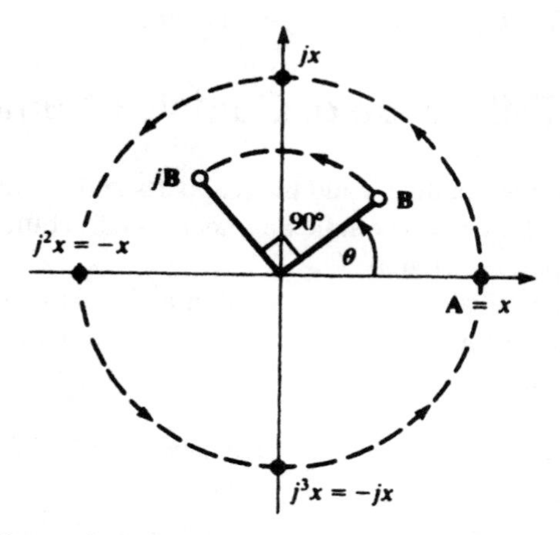

Figure A-2 Rotating complex numbers by j

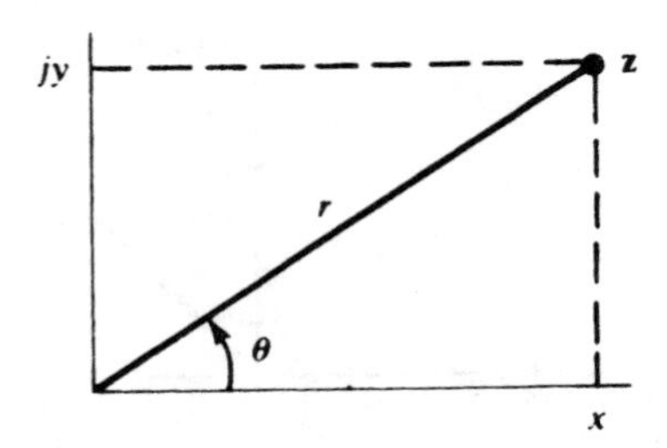

Figure A-3 Trigonometric-form for a complex number

In Figure A-3 $x = r\cos\theta$, $y = r\cos\theta$, and the complex number $\mathbf{z}$ can be written in *trigonometric form*

$$\mathbf{z} = x + jy = r(\cos\theta + j\sin\theta)$$

where r is the *modulus* or the *absolute value* (the notation $r = |\mathbf{z}|$ is common), given by $r = \sqrt{x^2 + y^2}$, and the angle $\theta = \tan^{-1}(y/x)$ is the *argument* of $\mathbf{z}$.

Euler's formula $e^{j\theta} = \cos\theta + j\sin\theta$ permits another representation of a complex number, called the *exponential form*: $\mathbf{z} = r(\cos\theta + j\sin\theta) = re^{j\theta}$

A third form, widely used in circuit analysis, is the *polar* or *Steinmetz form*, $\mathbf{z} = r\angle\theta$, where θ is usually in degrees.

Sum and Difference of Complex Numbers

To add two complex numbers add the real parts and the imaginary parts separately. To subtract two complex numbers, subtract the real parts and the imaginary parts separately.

From the practical standpoint, addition and subtraction of complex numbers can be performed conveniently only when both numbers are in rectangular form.

Example A.1 Find $\mathbf{z}_1 + \mathbf{z}_2$ and $\mathbf{z}_2 - \mathbf{z}_1$ where $\mathbf{z}_1 = 5 - j2$ and $\mathbf{z}_2 = -3 - j8$.

Solution:

$$\mathbf{z}_1 + \mathbf{z}_2 = (5 - 3) + j(-2 - 8) = 2 - j10$$

$$\mathbf{z}_2 - \mathbf{z}_1 = (-3 - 5) + j(-8 + 2) = -8 - j6$$

Multiplication of Complex Numbers

The product of two complex numbers when both are in exponential form follows directly from the laws of exponents.

$$\mathbf{z}_1\mathbf{z}_2 = (r_1 e^{j\theta_1})(r_2 e^{j\theta_2}) = r_1 r_2 e^{j(\theta_1+\theta_2)}$$

The polar product is evident from reference to the exponential form.

$$\mathbf{z}_1\mathbf{z}_2 = (r_1\angle\theta_1)(r_2\angle\theta_2) = r_1 r_2\angle(\theta_1+\theta_2)$$

The rectangular product can be found by treating the two complex numbers as binomials.

$$\begin{aligned}\mathbf{z}_1\mathbf{z}_2 &= (x_1 + jy_1)(x_2 + jy_2) = x_1x_2 + jx_1y_2 + jy_1x_2 + j^2 y_1y_2 \\ &= (x_1x_2 - y_1y_2) + j(x_1y_2 + y_1x_2)\end{aligned}$$

Example A.2 If $\mathbf{z}_1 = 5e^{j\pi/3}$ and $\mathbf{z}_2 = 2e^{-j\pi/6}$, then

$$\mathbf{z}_1\mathbf{z}_2 = (5e^{j\pi/3})(2e^{-j\pi/6}) = 10e^{j\pi/6}$$

Example A.3 If $\mathbf{z}_1 = 2\angle 30°$ and $\mathbf{z}_2 = 5\angle -45°$, then

$$\mathbf{z}_1\mathbf{z}_2 = (2\angle 30°)(5\angle -45°) = 10\angle -15°$$

Example A.4 $\mathbf{z}_2 = 2 + j3$ and $\mathbf{z}_2 = -1 - j3$, then

$$\mathbf{z}_1\mathbf{z}_2 = (2 + j3)(-1 - j3) = 7 - j9$$

Division of Complex Numbers

For two complex numbers in exponential form, the quotient follows directly from the laws of exponents.

$$\frac{\mathbf{z}_1}{\mathbf{z}_2} = \frac{r_1 e^{j\theta_1}}{r_2 e^{j\theta_2}} = \frac{r_1}{r_2} e^{j(\theta_1 - \theta_2)}$$

Again, the polar or Steinmetz form of division is evident from reference to the exponential form

$$\frac{\mathbf{z}_1}{\mathbf{z}_2} = \frac{r_1 \angle \theta_1}{r_2 \angle \theta_2} = \frac{r_1}{r_2} \angle(\theta_1 - \theta_2)$$

Division of two complex numbers in the rectangular form is performed by multiplying the numerator and denominator by the *conjugate* of the denominator (see section below for definition of the conjugate of a complex number).

$$\frac{\mathbf{z}_1}{\mathbf{z}_2} = \left(\frac{x_1 + jy_1}{x_2 + jy_2}\right)\left(\frac{x_2 - jy_2}{x_2 - jy_2}\right) = \frac{(x_1 x_2 + y_1 y_2) + j(y_1 x_2 - y_2 x_1)}{x_2^2 + y_2^2}$$

$$= \frac{x_1 x_2 + y_1 y_2}{x_2^2 + y_2^2} + j\frac{y_1 x_2 - y_2 x_1}{x_2^2 + y_2^2}$$

Example A.5 Given $\mathbf{z}_1 = 4e^{j\pi/3}$ and $\mathbf{z}_2 = 2e^{j\pi/6}$, then

$$\frac{\mathbf{z}_1}{\mathbf{z}_2} = \frac{4e^{j\pi/3}}{2e^{j\pi/6}} = 2e^{j\pi/6}$$

Example A.6 Given $\mathbf{z}_1 = 8\angle{-30^\circ}$ and $\mathbf{z}_2 = 2\angle{-60^\circ}$, then

$$\frac{\mathbf{z}_1}{\mathbf{z}_2} = \frac{8\angle -30^\circ}{2\angle -60^\circ} = 4\angle 30^\circ$$

Example A.7 Given $\mathbf{z}_1 = 4 - j5$ and $\mathbf{z}_2 = 1 + j2$, then

$$\frac{\mathbf{z}_1}{\mathbf{z}_2} = \left(\frac{4-j5}{1+j2}\right)\left(\frac{1-j2}{1-j2}\right) = -\frac{6}{5} - j\frac{13}{5}$$

Conjugate of a Complex Number

The *conjugate* of the complex number $\mathbf{z} = x + jy$ is the complex number $\mathbf{z}^* = x - jy$. Thus,

$$\operatorname{Re}\mathbf{z} = \frac{\mathbf{z}+\mathbf{z}^*}{2} \qquad \operatorname{Im}\mathbf{z} = \frac{\mathbf{z}-\mathbf{z}^*}{2j} \qquad |\mathbf{z}| = \sqrt{\mathbf{z}\mathbf{z}^*}$$

In the complex plane, the points $\mathbf{z}$ and $\mathbf{z}^*$ are mirror images in the axis of reals.

In exponential form, $\mathbf{z} = re^{j\theta}$ and $\mathbf{z}^* = re^{-j\theta}$.

In polar form, following the exponential form, $\mathbf{z} = r\angle\theta$, $\mathbf{z}^* = r\angle-\theta$.

In trigonometric form, $\mathbf{z} = r(\cos\theta + j\sin\theta)$, $\mathbf{z}^* = r(\cos\theta - j\sin\theta)$.

Conjugation has the following useful properties:

(i) $(\mathbf{z}^*)^* = \mathbf{z}$

(ii) $(\mathbf{z}_1 \pm \mathbf{z}_2)^* = \mathbf{z}_1^* \pm \mathbf{z}_2^*$

(iii) $(\mathbf{z}_1\mathbf{z}_2)^* = \mathbf{z}_1^*\mathbf{z}_2^*$

(iv) $\left(\dfrac{\mathbf{z}_1}{\mathbf{z}_2}\right)^* = \dfrac{\mathbf{z}_1^*}{\mathbf{z}_2^*}$

Index